Yanina Alejandra Corrada

Rehabilitation in the Canine and Feline Daily Clinic

Yanina Alejandra Corrada

Rehabilitation in the Canine and Feline Daily Clinic

Techniques and Practical Methodology

ScienciaScripts

This book is a translation from the original published under ISBN 978-613-9-46800-3.

Publisher:
Sciencia Scripts
is a trademark of
Dodo Books Indian Ocean Ltd. and OmniScriptum S.R.L publishing group

120 High Road, East Finchley, London, N2 9ED, United Kingdom
Str. Armeneasca 28/1, office 1, Chisinau MD-2012, Republic of Moldova, Europe
Printed at: see last page
ISBN: 978-620-6-64480-4

INDEX

PREFACE

The only impostole

It is that which is not attempted *Anonymously*

It was not until the second half of the 20th century that rehabilitation and physiotherapy therapies in small species gradually began to gain momentum. In spite of this, there are still very few reports in both canines and felines, and their use in the daily practice of clinical veterinarians is quite limited, perhaps due to the scarce information on the available techniques.

Life in cities implies a higher risk of accidents in pets, which have been increasing in recent years. On the other hand, the prolongation of life expectancy in canines and felines has led to a higher incidence of degenerative conditions resulting from aging, such as osteoarthritis, among others. Fortunately, the advance in the sophistication of the available complementary diagnostic methods, such as magnetic resonance imaging, has gradually allowed a greater certainty in the diagnoses and therapeutics used. All this has generated the need for new therapeutic alternatives complementary to traditional treatments, and thus physiotherapy is gradually becoming an important integral part of the treatment. Thus, places that usually work with high casuistry of traumatological and neurological cases are benefiting from the advantages of having physiotherapy for early recovery of their patients. For the professional, the ultimate goal is usually the rehabilitation of the patient, which implies an adequate pain control, the achievement of the maximum functionality of the injured area, a correct control and prevention of secondary lesions, with a return to daily activity with the maximum possible independence. Regarding the owner, their goal most of the time is simply that their pet has the best quality of life. And it is our obligation as professionals, to exercise our profession in pursuit of this.

Thus arose the need to write our first physiotherapy book, and with it the continuation through this issue. We hope its reading will be useful to the professional and advanced student of the career to develop their interest in the specialty of physiotherapy and canine and feline rehabilitation.

ACKNOWLEDGMENTS

To the *authors* of each capital for their valuable contribution in the preparation of this book.
book

To the *Owners* for entrusting us with the health of their pets

To our *Patients* for reminding us of the beautiful profession we have chosen: *Being*

Veterinarians.

To my *Marna* and my *brother* for their unconditional support
To my *children,* for teaching me the value of simplicity

REHABILITATION IN THE DAILY CANINE AND FELINE PRACTICE

Yanina Alejandra Corrada

Doctor of Veterinary Medicine

Specialist in Small Animal Clinic CONICET Researcher

The editor, Dr. Yanina Alejandra Corrada, created in the Faculty of Veterinary Medicine of the National University of La Plata, the Area of Veterinary Physiotherapy, thus promoting various studies on the subject as well as training of human resources.

CANINE AND FELINE PHYSIOTHERAPY AND REHABILITATION TECHNIQUES

Alvarez FG[1,2] ; YA Corrada[2,3,4]

[1]Scholarship holder Consejo de Investigaciones Cientificas y Técnicas (CONICET);[2] Physiotherapy Laboratory - LAFIVET;[3] Chair of Small Animal Clinic; Faculty of Veterinary Sciences, National University of La Plata; 3 CONICET researcher.

SUMMARY

Physiotherapy has many technologically advanced tools, among which are mainly thermotherapy, massage therapy, kinesitherapy and electrotherapy.

Thermotherapy is the application of heat and cold as therapeutic means. The application of cold in general induces vasoconstriction (decrease of local blood flow), and helps in the reduction of edema, pain and muscular spasm, being contraindicated in animals with circulatory problems, open wounds, ischemic lesions or with alterations of sensibility. The application of heat is indicated after the resolution of acute inflammation because it produces local vasodilatation and pain reduction, being inadvisable its use in cases of acute local inflammation, tumors, open wounds, hematomas, serious circulation problems and absence of sensitivity. *Massage therapy* is the manipulation of soft body tissues through the use of the hands. It has effects on the skin, skeletal musculature, joints, circulatory system and nervous system. *Kinesitherapy* is the part of physical medicine that uses movement as a treatment. The objectives pursued by kinesitherapy are: *preventive*, maintaining the extensibility potential of the musculo-tendinous structures avoiding the loss of the body scheme; *palliative*, attenuating the secondary effects of the disease; *curative*, recovering the joint mobility, increasing the trophism and muscular power and improving the stimuli informing the movement. Electrotherapy is very efficient in the treatment of orthopedic and neurological affections, but cannot be used on eyes and glands, being contraindicated in open wounds, inflamed regions, infections, tumors, in the treatment of occasional pain and in anesthetized areas.

INTRODUCTION

Physiotherapy has many technologically advanced tools, among which are mainly thermotherapy, massage therapy, kinesitherapy and electrotherapy.

As in any other specialty, a complete assessment of the patient is essential in order to be able to propose a correct treatment.

The success of a rehabilitation treatment depends on good cooperation and performance between the different parties involved:

- The veterinary professional must be competent.

- The family must be involved.

- The pet must cooperate

Thermotherapy

Thermotherapy is the application of heat and cold as therapeutic means.

The application of cold as a therapeutic agent (cryotherapy), is related to the neurovegetative system that produces reactions of the vascular system with decreased circulating blood volume, acceleration of voluntary movements and reflexes, and decreased pain perception. Thus, the application of cold in general induces vasoconstriction (decrease of local blood flow), and helps in the reduction of edema, pain and muscle spasm. It is generally used in acute episodes, as well as in the immediate post-operative period, as prevention after physical exercise, trauma, arthritis, myositis, and tendonitis among others.

It can be done through the application of refrigerants or ice wrapped in a towel (in order to avoid direct contact with the skin). The cold compress is applied directly on the treated area, in general sessions of 10-15 minutes are performed. The interval between sessions varies according to the case, for example: for most of the postoperative indications it should be done every 3 or 4 hours while in more serious cases the cold applications could be repeated every 2 hours.

In case of using commercial refrigerants, special care must be taken to prevent the dog from chewing them as they may contain a gel that is toxic to animals.

Local effects during acute inflammation:

Decreased local metabolism and histamine release, among others, at the origin of vasodilatation and edema.

Stimulation of cutaneous receptors, sensitive to acute pain and cold. These are type III (myelinated) and type IV (unmyelinated) thermoreceptors. This causes: decrease in the speed of conduction of the nerve impulse and increase in the refractory period of stimulation, with the consequent reduction of pain sensation; decrease in muscle spasm, interfering on the neuromuscular spindle.

Systemic effects during acute inflammation:

•   Hypothalamic and autonomic nervous system response with generalized vasoconstriction at cutaneous level, with modification of heart and respiratory rate, and appearance of tremors.

Cryotherapy is contraindicated in animals with circulatory problems, open wounds, ischemic lesions, with alterations of sensitivity, malignant tissues, scar tissue (poor blood supply), or in areas where radiotherapy or other ionizing radiation has been applied in the last 6 months.

The application of heat is indicated after the resolution of acute inflammation because it produces local vasodilatation, decreased pain, increased oxygen supply and tissue metabolism, increased collagen viscosity, relaxation of muscle spasm, increased capillary permeability, and improved healing. Local heat is widely used in cases of chronic arthritis, as well as in cases of stiffness and muscle tension, leaving the area conducive to receive therapeutic massage or any other technique.

In heat therapy there are two types of heating:

• Superficial: penetrates to a depth of approximately 1o2 cm

• Deep: penetrates to a depth of 4 cm

Superficial forms include hot compresses, hot water bottles, baths, heating pads, and hot air dryers among others. This type of therapy is very useful in subacute and chronic problems to relieve pain, it can also be used prior to passive movements, stretching or exercises.

Deep heating is achieved through the application of electrotherapy and/or therapeutic ultrasound, which should only be performed by competent and qualified personnel due to the inherent potential dangers such as burns, tissue destruction, blood cell stasis and endothelial damage.

The heat can be applied by means of hot towels, compresses with hot water bags, among others. The time of each application is about 15 to 20 minutes, and the total duration of treatment as well as the frequency should be defined according to the evolution of each clinical case. As explained above, with these home methods the penetration achieved is approximately 1 cm. In case of needing a greater penetration, ultrasound should be used, with which 1 to 3cm deep penetration is achieved.

Ultrasound is widely used to access deeper tissues. As the waves are absorbed by the muscle, the mechanical energy is transformed into heat. In physiotherapy and rehabilitation, waves of frequencies ranging from 0.5 to 5 MHz are used, although the equipment is usually 1 MHz (low frequency) to 3.3 MHz (high frequency). The frequency to be used will depend on the depth of the tissue to be treated. If the desired penetration depth is between 0 and 3cm (bone areas) the frequency to be used would be 3MHz, while if it is between 2 and 5 cm (deeper tissues) a lower frequency of 1 MHz would be used. The intensity corresponds to the rate of energy distributed per unit area (W/cm).[2]

Two ultrasound transmission modes can be used: continuous or pulsed mode.

pulsed mode. In pulsed mode it is possible to choose the duration of the transmission and pause period, which allows the use of higher intensities to reach the desired temperature levels without causing residual damage or altering the permeability of the cells and improving their metabolism.

It should be taken into account that in order to apply ultrasound correctly, a trichotomy of the area should be performed. A factor to take into account when applying this therapy is that it is always advisable to shave the hair of the treated area, this is because the transmitted energy is absorbed by the hair layer of the animal due to its high protein content. A thick layer of gel does not improve this situation.

Physiological effects:

Temperature has a direct influence on the speed of chemical and biochemical reactions, in particular on enzyme activity, but also on the interaction of molecules. For example, it decreases the binding of 02 to hemoglobin, with the consequent greater availability of the latter. Temperature also modifies the assimilation of 02 at tissue level. These conditions make it possible to accelerate tissue repair.

Vasodilatation at cutaneous level, by means of three mechanisms:

-    Direct release of chemical mediators such as histamine and prostaglandins.

-    By stimulation of cutaneous thermoreceptors connected in turn to the cutaneous blood vessel walls. Their stimulation induces the release of bradykinin and relaxation of the smooth muscle layer of the blood vessel walls.

- Reduction of sympathetic nervous system activity, reducing local smooth muscle contraction.

Modification of nerve impulse conduction velocity and refractory period at the level of sensory and motor nerves. For each additional degree of temperature, the conduction velocity of the nerve impulse increases 2 m/s.

Modification of the excitability threshold of the neuromuscular spindle with consequent muscle relaxation.

Pain control. Three different mechanisms are recognized to increase its threshold:

- Stimulation of cutaneous thermoreceptors

- Vasodilatation reduces ischemia at tissue level, with consequent less stimulation of nociceptors.

- Reduction of muscle spasm that relieves pressure on the walls of blood vessels

As explained above, it is known that the application of ultrasound stimulates or favors the normal sequence of events that occur during healing and tissue repair, thus increasing its effectiveness. It also influences scar tissue remodeling by favoring the appropriate orientation of the neoformed collagen fibers, thus increasing the tensile strength and favoring the passive mobility of the scar (Nussbaum, 1998). More recent studies have also identified the benefits of using low intensity pulsed ultrasound for bone fractures that are healing normally, delayed fractures or nonunions (Warden et al., 1999; Tis et al., 2002; Sakurakichi et al., 2004).

- Similarly, heat increases the elasticity of connective tissue in tendons, ligaments and joint capsule.

The application of heat is contraindicated in cases of acute local inflammation, tumors, open wounds, hematomas, severe circulation problems and lack of sensitivity.

Massage therapy

Massage therapy is the manipulation of soft body tissues through the use of the hands. It has effects on the skin, skeletal musculature, joints, circulatory system and nervous system. It is widely used in both human and veterinary physiotherapy. In general, animals respond well to this procedure, which certifies that it is a conforming technique for the patient. When the tissue is subjected to a force, the collagen responds with elongation. As long as that force remains within the limits of the basal region, the tissue is able to stretch without irreversible damage, so the massage has to act within those limits. In the case that the applied force is maintained over time or increases in intensity, microdefects are created, so that the fibers will present a new length, progressively losing elasticity and becoming more rigid (linear or plastic region), until reaching the maximum point of rupture. Therefore, care must be taken when applying the massage as it could damage the internal structure of the connective tissue, causing loss of functionality. Whether the animal is large or small, it is necessary to place it in a calm and comfortable environment to achieve relaxation during the treatment. By means of the massage, muscular tension is reduced, circulation is improved and muscle recovery is accelerated.

Massage therapy is indicated to relieve pain and muscle tension, as well as for joint problems, adhesions, muscle tone regulation, as well as prevention and recovery from injuries.

Massage therapy is contraindicated when there is local inflammation, infections, areas with dermatitis, tumors, hyperthermia, cardiac insufficiency and circulatory disturbances.

Massage should be performed with caution in: acute neural pathologies (e.g. disc disease) where stimulation may be uncomfortable for the animal or in certain degenerative pathologies (e.g. arthritis) where pressure may be uncomfortable.

The massage techniques that can be used are varied. The most commonly used in veterinary medicine are the following:

1.   Superficial Gliding: the animal should be in station, and the professional performing the procedure should be positioned next to or behind the animal. The hands should be placed gently on the animal, with movements in favor of the hair, starting from the neck, and continuing on the thorax, abdomen, hips and ending on the limbs. It is used to relax the animal, manipulate the superficial muscles and improve blood and lymphatic circulation. This technique also allows the animal to get used to the contact and is useful to start and end a massage session.

2.   Friction: it can be performed all over the animal's body or only in one area, reaching the deepest tissues, or superficially. It helps to increase the blood flow, promote the elimination of toxins and collaborates to decrease in some cases the adhesions, as well as to reduce the tensions. The technique is performed with the fingertips or the fingertips, sliding carefully forward or backward, and in the opposite direction of the muscle fibers. The pressure should be increased according to the needs. Friction is contraindicated in acute inflammations and infections.

3.   Shake: It is used to relax the muscles and as a complement to other exercises. It takes a set of muscles moving them with small jerks forward and backward.

4.   Percussion: in general terms, these are a series of techniques in which the hands strike the body. They are used to relax the muscles, improve circulation, and are particularly useful for patients with hypotonia or in the postoperative care of a patient's chest. It is performed with very gentle strokes with the palm of the hand cupped.

5.   Pressure technique (kneading): consists in compressing and releasing the muscles and subcutaneous tissues in a rhythmic way to improve circulation and lymphatic flow.

Kinesitherapy

Kinesitherapy is the part of physical medicine that uses movement as a treatment. The objectives of kinesitherapy are: *preventive*, maintaining the extensibility potential of the musculo-tendinous structures avoiding the loss of the body scheme; *palliative*, attenuating the side effects of the disease; *curative*, recovering joint mobility, increasing trophism and muscle power and improving the stimuli informing the movement.

Among them therapeutic exercises used in kinesitherapy can be considered:

a)   Passive therapeutic exercises: seeks the movement of body segments by means of a force independent of the patient's neuromuscular units. It is used to maintain normal range of motion, improve flexibility of muscles, tendons and ligaments, as well as flexion and extension of joints, neuromuscular structure and function. Passive movements will not prevent muscle atrophy. It is usually advisable to start passive movements as soon as possible after the injury, the planned movements should be comfortable for the patient. It is a technique widely used in dogs with ruptured intervertebral discs. As its name indicates, the exercise is passive, i.e. the movement must be performed entirely by the physiotherapist. Because of this, a good knowledge of both the anatomy and physiology of the animal is necessary. The most frequent indications include: animals in decubitus, patients with chronic degenerative pathologies and during recovery from neurological diseases, among others.

b)   Range of motion (ROM) exercise: very important exercise in young animals. It is ideal for dogs that have undergone joint surgery. It is performed by extending and flexing the limb, always in a gentle manner so that it is comfortable for the animal. In some cases, rotation movements can be performed.

c) Stretching exercises: most of the time it is used together with the previous one, mainly in cases of muscle stiffness and decreased range of motion (ROM). It aims to elongate and realign the soft tissue and collagen. It is performed through the mobilization of a joint. They can be more effective if performed after the application of massage or hot packs and after exercise. There are several different types of stretching, the most commonly used in the small animal clinic are static stretching and prolonged mechanical stretching. Both types are characterized by being low intensity exercises, where the force applied is low, the difference between one and the other lies in the time in which the force is maintained, being more prolonged in the second (from 20min to several hours). In addition, prolonged mechanical stretching requires the use of splints.

d) Flexor reflex exercises: is used in patients with neurological problems. It is performed by gently pinching the hairs between the fingers, and in response the patient should flex the stimulated limb. That is to say that its objective is muscle contraction.

e) Bicycle movements: to perform this exercise it is better to place the animal in lateral decubitus. It is widely used, in association with ROM, in animals with neurological problems, and in animals with difficulties to support their weight on their limbs. If it is performed with the patient in station, you can help him by placing a hand on his side.

Assisted Kinesitherapy is applied when the animal is able to stand by supporting its own weight.

a) Stationary maintenance is widely used in patients with neurological disorders, as well as hip conditions. It helps to increase strength, endurance and neuromuscular sensitivity. As the name implies, the animal actively participates in these exercises, always with the help of the clinician. Sometimes a towel or strap may be used under the abdomen to help the patient walk. The movements should be slow, always respecting the animal's time.

b)   The balance board is used to strengthen balance. The animal is placed on an unstable board, it always helps to avoid falling.

c)   Therapeutic ball: used in animals with support problems. The animal is placed in station on a ball that allows only the support of the forelimbs and hind limbs, making movements forwards and backwards.

d)   Cervical flexion and extension, and lateralities: aims to improve the mobility of the animal. The movement of the head is stimulated both laterally and ventrodorsally. It is useful to use rewards so that the animal collaborates with the exercises.

Active Kinesitherapy aims to restore the patient's strength and locomotion, as well as improve cardiorespiratory capacity and pain relief. It requires that the animal must first be able to support its own weight before beginning the exercises.

The types of therapeutic exercises that can be performed can be grouped into 4 general categories:

•   Strengthening (power): exercises to improve strength increase the cross-sectional area of the muscle. Strengthening exercises include activities such as running, stair climbing and descending, weight pulling, "dancing", wheelbarrow and swimming.

•   Endurance (resistance): this type of exercise generates in the muscle an increase in vascularization, which translates into a greater arrival of oxygen to the organ. In turn, at the systemic level, it leads to a decrease in heart rate at rest and an increase in stroke volume, which together result in a decrease in blood pressure. Resistance exercises include jogging, swimming, treadmill activity and sled pulling.

•   Flexibility (agility): flexibility is particularly important in cats and sporting dogs, although all animals require good flexibility to avoid injury. Exercises that strengthen this quality are those that make the animal reach or stretch for something. These activities include crawling underneath, stepping over obstacles, climbing stairs, and interlocking exercise.

• Balance and proprioception: proprioception decreases with age, and can also be affected by injury or surgery. Balance and proprioception exercises are activities that require a quick response to changes in the support surface. These include wobble boards, balancing on padding, trampolines, dancing, standing on a gym ball, walking in circles, weaving and walking over obstacles.

The type of exercise to be selected varies according to the clinical picture and health conditions of the animal. The following is a simple description of some of the exercises mentioned above:

a. Walks: should be carried out with great care, always respecting the animal's condition, without demanding or straining it too much. It can be started with 2 to 5 minutes each time, 2 to 3 times per day, and then increase it. Attention should be paid to the appearance of claudication, in case it happens, decrease the time and frequency.

b. Climbing and descending stairs: good exercise to work on joint extension. It should be performed when the animal is able to walk safely. The exercise should be done slowly and always well monitored.

c. Dance: it is performed by lifting the animal's forelimbs and moving it forward and backward. It is used to strengthen the muscles of the hind limbs.

d. Cart: used to strengthen the musculature of the forelimbs. It is performed by raising the hind limbs and making the animal walk with the forelimbs.

e. Stand up/sit up: used especially in cases of dysplasia and hind limb disorders.

f. Treadmill: it is widely used, since the walk can be adapted to the dog's conditions.

g. Hydrotherapy: it is considered one of the best methods to achieve muscle strengthening, increase mobility, improve body condition, blood circulation and cardiorespiratory capacity.

Electrotherapy

Electricity applied to the body for therapeutic purposes constitutes *electrotherapy*, being very efficient in the treatment of orthopedic and neurological conditions. It is also very useful in cases of muscular atrophy resulting from neurological situations, muscular ruptures and tears, fracture consolidation, circulatory problems, as well as in the management of acute and chronic pain resulting from neurological conditions such as cauda equina and Wobbler's syndrome.

Electrotherapy cannot be used on eyes and glands, being contraindicated in open wounds, inflamed regions, infections, tumors, in the treatment of occasional pain and in anesthetized areas.

In order to place the electrodes correctly and improve conductivity, it is necessary to perform the correct trichotomy of the area to be treated. The electrodes should be placed opposite to the hair growth, and once placed can be fixed with a girdle or held by the physiotherapist. The stimulation parameters such as intensity, frequency and duration should be adjusted to each patient according to the type and severity of the lesion as well as the response to treatment. In general, when the case is acute, the intensity used is usually low.

The animal's response should be evaluated at the end of each session.

With the use of a TENS (transcutaneous neuromuscular electrical stimulator) low frequencies (1 to 3 Hz) and medium frequencies (80 to 150 Hz) are reached, which stimulate peripheral nerves and muscles and cause analgesic effects through synchronous depolarization of the nerve fibers. In this way, acute and chronic pain is reduced. It should be noted that the electrodes should be placed on the affected joint or along the dorsal muscles, transversally or longitudinally.

Physiotherapy in cats:

The principles and techniques of physical therapy described in this chapter are applicable to all species, but the feline patient requires a special mention, this is because in general terms it is a less tolerable patient to the usual management that is performed in physical therapy sessions. It is for this reason that when treating feline patients with physical therapy it is necessary to modify the techniques, therefore to obtain satisfactory results and not to fail in the therapy of cats it is advisable to carry out shorter treatment sessions, to introduce the exercises in a more gradual way and to use familiar items for the cat, such as toys.

Manual therapies and electrotherapy can be used successfully in cats. Many of the exercises described above, such as "dancing" and "wheelbarrow", can be easily adapted. While other exercises such as leash walking and hydrotherapy are more difficult to perform, therefore the benefits of the latter two therapies will be reserved for a few individuals.

Finally, it is also advisable to take advantage of the feline patient's hunting instinct to perform activities that help improve strength, joint range of motion and balance. Among the exercises to encourage feline hunting, we can mention the toys dragged on the floor and the laser flashlight, among others.

CONCLUSIONS

In veterinary medicine, physiotherapy was initially used with exercises and techniques extrapolated from human medicine on athletic horses or those with orthopedic problems. With the passing of time and the advance of scientific studies, these techniques underwent the necessary adaptations that allowed not only a better response in the patient, but also made possible their use in small animals. Likewise, the incorporation of new equipment was expanding the therapeutic possibilities in the different pathological conditions. Physiotherapy has many technologically advanced tools, among which are mainly thermotherapy, massage therapy, kinesitherapy and electrotherapy. Each of them has its specific indications and contraindications, which must be known in order to optimize the early recovery of the patient.

BIBLIOGRAPHY

1. Bockstahler B, Levine D, Millis D. Essential Facts of Physiotherapy in Dogs and Cats. Rehabilitation and Pain Management. 1st Edition. Babenhausen, Germany. 2004

2. Buchner HHF, Schildboeck U. Physiotherapy applied to the horse: a review. Equine Veterinary Journal. 2006; 574-580

3. Clark B. Physical Rehabilitation in Small Animal Orthopedic Patients. Veterinary Medicine. 2001; 234-246

4. Douglas S; Pippi N; Costa F; Vescovi L; Conti L; Weiss A. 2008. O ultra-som terapèutic de 1 MHz, na dose de 0,5 W cm-2, sobre o tecido ósseo de càes avaliado por densitometria òptica em imagens radiograficas. Ciència Rural. 38: 8. 2225-2231

5. Downer A. Underwater Exercise for Animals. Modern Veterinary Practice. 1979; 115-119

6. Downer A. Physical Therapy for Animals. Selected Techniques. Springfield. 1978

7. Haltrech H. Physical Therapy. Can Vet J. 2000; 41:573-574.

8. Jaffery N. Spinal Cord Injury in Small Animals. Current and Future Options for Therapy. TheVeterinary Record. 1999; 14:183-190

9. Marsolais G, Devorak G. Effects of postoperative rehabilitation on limb function after cranial cruciate ligament repair in dogs. JAVMA. 2002; 9:

13251330

10. Mills L, Taylor M. Rehabilitation and Physical Therapy. Veterinary Clinics of North Am Small Animal Practice. 2005.

11. Pena A. Low frequency electrotherapy (TENS). Fundamentals, therapeutic possibilities and indications. 1992; 3: 143-154

12. Prentice WE. 1990. Sports Medicine. Therapeutic Techniques. St Louis Missouri. Ed Mosby. 1990

13. Ruiz Perez M. Canine Rehabilitation and Physical Therapy. Ed Intermedica. 2011

14. Santoscoy Mejia EC. In: Orthopedics, Neurology and Rehabilitation in Small Species. Dogs and Cats. 2008

15. Server R; Hury L. 2016. Introduction to the World of Rehabilitation - Physiotherapy. Asis Formation.

16. Speciale J, Fingeroth J. Use of Physiatry as the Sole Treatment Three Paretic or Paralyzed Dogs with Chronic Compressive Conditions of the Cauda Portion of the Cervical Spinal Cord. JAVMA. 2000; 217:221

HISTORY OF PHYSIOTHERAPY AND REHABILITATION VETERINARY

Corrada YA[1,2,3]

[1]Chair of Small Animal Clinic; ^Physiotherapy Laboratory -

LAFIVET' Faculty of Veterinary Sciences' National University of La Plata;

[3]CONICEт

SUMMARY

The function of physiotherapy lies in achieving better conditions' both for trauma and post-surgical recovery' as well as for pain reduction' muscle strengthening and locomotion improvements. Sometimes' it aims to achieve the maintenance or decrease of weight in overweight animals' and with this help to prevent the presentation of secondary ailments to overweight and obesity. For the World Confederation of Physical Therapy (WCPT) in 1995 defined it as "*Specialty of the health area whose main purpose is the optimal promotion of health and function, including the generation and application of scientific principles in the process of evaluation, diagnosis and functional prognosis and physiotherapeutic intervention, to prevent or remedy functional limitations and disabilities related to movement*". *Rehabilitation* is currently defined as the global and dynamic process that is carried out aiming at the physical recovery of an animal suffering from any impairment or difficulty and seeking to achieve a better quality of life and physical well-being. *Physiotherapy* is currently defined as the science applied to the study, diagnosis, prevention and treatment of organ and system dysfunctions. To perform it is necessary to have extensive knowledge of the structures and functions of the body. Among the conditions that benefit from the use of physiotherapy and rehabilitation are fractures, joint pathologies, muscle-tendon pathologies, neurological patients.

INTRODUCTION

The main objective of animal physiotherapy is to improve the quality of life of the treated animals. Its function is to achieve better conditions, both for trauma and post-surgical recovery, as well as to reduce pain, strengthen muscles and improve locomotion. Pain relief is necessary because it can bring with it associated undesirable situations such as poor appetite, weight loss and poor quality of life in general. It is also used to minimize muscle atrophy and achieve improvements in tendons, ligaments, cartilage and bones. In post-surgical rehabilitation, physiotherapy acts with the objective of minimizing possible sequelae and accelerating recovery, consequently reducing the use of medication drugs.

Occasionally, physiotherapy aims to achieve weight maintenance or weight loss in overweight animals, thus helping to prevent the development of ailments secondary to overweight and obesity.

Evolution

Physiotherapy has been practiced for healing purposes since ancient times by the Egyptians, Greeks, Romans, Chinese, among others, however, it was not until after the Second World War when rehabilitation and physiotherapeutic techniques were developed as medical specialties. At that time they acquired great importance worldwide by demonstrating that invalids subjected to appropriate physiotherapeutic treatments achieved unsuspected degrees of recovery. The term "Physiotherapy" comes from the union of the Greek words *"Physis"* which means *"Nature"* and *"Therapehia"* which means *"Treatment"*. This means that etymologically physiotherapy means *"Treatment by Nature"*, although today its meaning is recognized as *"Treatment by Physical Agents"*. The change of the etymological sense is consolidated from the World Health Organization that in 1958 defines Physiotherapy as *"The art and science of treatment by means of therapeutic exercise, heat, cold, light, water, massage and electricity"*. This definition is fundamental in the history of physiotherapy since it mentions the therapeutic exercise and therefore incorporates an important element that the physiotherapist has as exercise from a therapeutic conception through *movement.*

Beyond being established as a relatively new discipline, physiotherapy currently has as many definitions as countries and associations where it is practiced. Therefore, the World Confederation of Physical Therapy (WCPT) in 1995 defined it as "*Specialty of the health area whose main purpose is the optimal promotion of health and function, including the generation and application of scientific principles in the process of evaluation, diagnosis and functional prognosis and physiotherapeutic intervention, to prevent or remedy functional limitations and disabilities related to movement*. In veterinary medicine, physiotherapy was initially used with exercises and techniques extrapolated from human medicine on athletic horses or those with orthopedic problems. With the passage of time and the advance of scientific studies, these techniques underwent the necessary adaptations that allowed not only a better response in the patient, but also made possible its use in small animals.

The pioneers in animal physiotherapy were Sir Charles Strong and Lord Louis Mountbatten, who put into practice their skills in massage therapy with racing horses, as well as to repair injuries of ponies destined for the sport of polo. Thus, the veterinary massage therapist became the physiotherapist for horses. In 1978 the techniques for canine physical therapy were described by Ann Downer, a therapist at Ohio National University. From then on, specialized texts began to be published giving guidelines for the application of the therapy.

Currently, racehorses and show horses are subjected to physiotherapeutic agents such as infrared rays, ultrasound, thermotherapy, hydrotherapy and acupuncture, thus pursuing the goal of significantly reducing recovery time in sports injuries and any other type of injury that affects locomotion. In small animals, the field of physical rehabilitation has much to offer to professionals, and its use has been especially focused on the alterations occurred in sporting dogs, as well as patients with orthopedic and neurological injuries. In the second half of the 20th century, rehabilitation therapy began to gain momentum in small species, although there are very few reports in dogs. However, despite the extensive history of physiotherapy in human medicine, the application of physiotherapeutic techniques in the daily practice of clinical veterinarians is limited, perhaps due to the lack of information regarding the techniques, as well as the scarce diffusion of this field in the veterinary medicine of the country.

Indications

With the passage of time, physiotherapy became a great ally of orthopedic surgeries. Thus, in the past, the recommendation was that the animals submitted to orthopedic interventions should be kept at rest, and when they began to exercise, it should be done very gradually. Nowadays this has changed, and some exercises are indicated to the owner so that the animal can recover quickly. This shows the importance, even if you are not a veterinary physiotherapist, of knowing the basic techniques of physiotherapy.

Among the most common conditions treated with the help of physiotherapy are arthritis, elbow dysplasia, coxofemoral dysplasia, tendinitis, bursitis, ligament injuries, pre and post-surgical and fracture healing. It can also be indicated for pain relief, acceleration of healing of some types of wounds, animal athletes, edema, blood and lymphatic circulation disorders, complications in the cardiorespiratory system and weight loss.

Among the specific objectives of physiotherapy are:

- Reduction of pain and inflammation

- Recovery of functionality

- Elimination of the cause of the dysfunction

- Prevention and reduction of muscular atrophy and injuries to cartilage, bones, tendons and ligaments.

- Improvement of the patient's general condition and, in particular, his cardiovascular endurance

- Reduction of edema

- Increased collagen production

- Tissue repair

Before describing the techniques themselves, it is important to know some basic concepts. *Rehabilitation is* currently defined as the global and dynamic process that is carried out in pursuit of the physical recovery of an animal with a deficiency or difficulty, and seeking to achieve a better quality of life and physical well-being. That is to say, rehabilitation is carried out through a large set of actions, being a very dynamic process in which the patient is subjected to a continuous assessment throughout the rehabilitation. Because of this, the treatment must be readapted according to the evolution of the treated animal.

In rehabilitation, non-invasive techniques are used, based on the knowledge of the physiological and pathophysiological characteristics of the tissues, and in particular of the locomotor and nervous system.

Some of the advantages offered by veterinary rehabilitation are:

- Improved functionality and quality of movement

- Pain control by reducing the administration of non-steroidal anti-inflammatory drugs (NSAIDs) to their minimum indispensable dose.

- Acceleration of the recovery process, with maximum control over possible secondary injuries.

- Improvement of the patient's quality of life

Any animal species, breed, sex or age is suitable to undergo rehabilitation treatment, although the equine and canine species, due to their idiosyncrasy, are the ones that best respond to this type of treatment.

In all cases, in its application, rehabilitation must take into account the individual and his or her conditions, as well as the evolution of his or her general state of health.

Physiotherapy is currently defined as the science applied to the study, diagnosis, prevention and treatment of organ and system dysfunctions.

To perform it is necessary to have a wide knowledge of the structures and functions of the organism. Physiotherapy studies, diagnoses, prevents and treats the disturbances referred to alterations of organs and systems of the animal, as well as the beneficial effects of the physical resources used such as thermotherapy, massage therapy, electrotherapy, kinesitherapy and hydrotherapy among others.

There are many cases in which the functional recovery time has been significantly reduced after an appropriate diagnosis and surgical treatment.

Conditions benefited by the use of physiotherapy and rehabilitation:

Fractures: Multiple elbow fractures, whether "Y" or "T", required after surgical stabilization a recovery period of sometimes several months, no matter how well the technique had been performed. Nowadays, this functional recovery time can be reduced to half or less, in most of the cases in which the patient can attend the joint rehabilitation sessions.

Joint pathology: In those patients who undergo surgery for a joint pathology, their owners can be offered the possibility of attending physiotherapy and rehabilitation sessions that the specialist considers necessary until the recovery of joint mobility. Some examples are the rupture of the anterior cruciate ligament in the knee (Figure 1), arthroplasty of the femoral head and neck in cases of Legg-Calvé-Perthes disease or hip dysplasia with joint pain, osteochondritis dissecans of the shoulder, or fragmentation of the medial coronoid process in cases of elbow dysplasia.

In cases of medial luxation of the patella in mini breeds, you can opt for a conservative treatment in grades I and II, so that the animal will perform a few sessions of strengthening of the quadriceps that, in most cases, allow not having to undergo surgery.

Figure 1: Anterior cruciate ligament knee surgery in Rottweiler

Pathologies of muscle-tendon origin: Another group of pathologies in which the appearance and development of physiotherapy and rehabilitation techniques have been decisive in their treatment are those of muscle-tendon origin, such as fibrosis-contracture of the gracilis muscle or of the semitendinosus muscle, calcification of the tendon of the biceps brachii muscle or of the supraspinatus muscle.

In patients with fibrosis-contracture of the gracilis muscle, not surgically intervened, it can be achieved that the process does not progress, that the muscular discomfort disappears and that the animal leads a practically normal life with its "different" gait. Other cases require surgery and subsequent physiotherapy and rehabilitation sessions.

Neurological patients: Among the patients who can benefit from these disciplines, neurological patients are the star group, both those who require surgical treatment and those who do not.

Neurological patients with surgical treatment

• Herniated discs: these patients deserve special attention, since this is one of the most frequent pathologies among neurological cases. Types:

Hansen type I: acute rupture of the annulus fibrosus in a momentary effort and extrusion of the usually degenerated nucleus pulposus.

• Hansen type II: the fibrous ring is gradually deformed and generally does not break, so there is no exit of disc material into the spinal canal and sometimes, depending on the spinal area, does not require surgical treatment.

Both situations have seen the recovery time shortened dramatically since the incorporation of physical therapy and rehabilitation in the early postoperative period, being even shorter when the neurological condition of the patient before entering the operating room was relatively satisfactory.

• Alterations of the atlanto-axial joint: animals of mini or *toy* breeds suffering from alterations of the atlanto-axial joint, either by pathology of the axis tooth, of the ligaments that stabilize it or both. Neurological patients with non-surgical treatment.

• Fibrocartilaginous thromboembolism: generally large or giant breeds suffering from fibrocartilaginous thromboembolism in which, if the initial pathology does not seriously injure the spinal cord, they recover in a very high percentage of cases and much faster.

• Meningitis or meningo-encephalitis, peripheral polyneuropathies, neuritis and polyneuritis: medium or small breeds that can suffer meningitis or meningo-encephalitis, peripheral polyneuropathies or neuritis and polyneuritis due to different causes that, once treated, have also seen their recovery period extraordinarily shortened. Examples are patients affected by *Toxoplasma and Neospora*.

CONCLUSIONS

Physiotherapy and rehabilitation, although they are very old medical specialties, have little development in current veterinary medicine. *Rehabilitation* is the global and dynamic process that is carried out pursuing the physical recovery of an animal with some deficiency or difficulty, and seeking to achieve a better quality of life and physical well-being. *Physiotherapy* is the science applied to the study, diagnosis, prevention and treatment of dysfunctions of organs and systems. To perform it is necessary to have extensive knowledge of the structures and functions of the body. In rehabilitation non-invasive techniques are used, based on the knowledge of the physiological and pathophysiological characteristics of tissues, and in particular of the locomotor system and nervous system. Some of the advantages offered by rehabilitation are the improvement in the functionality and quality of movements, as well as pain control by reducing the administration of non-NSAID anti-inflammatory drugs to their minimum indispensable dose, acceleration of the recovery process, with maximum control over possible secondary injuries, and improvement of the patient's quality of life. However, beyond the extensive history of physiotherapy in human medicine, the application of physiotherapeutic techniques in the daily practice of clinical veterinarians is not yet a reality.

The information about the techniques is limited, perhaps due to the lack of information about them, as well as to the scarce diffusion of this field in the veterinary medicine of the country. That is why it is a developing specialty with much to offer to veterinary professionals.

BIBLIOGRAPHY .

1.   Bockstahler B, Levine D, Millis D. Essential Facts of Physiotherapy in Dogs and Cats. Rehabilitation and Pain Management. 1st Edition. Babenhausen, Germany. 2004

2. Bromiley MW. Physical Therapy for the Equine Back. The Veterinary Clinics of North America. Equine Practice. 1999; 15(1):223-246

3. Buchner HHF, Schildboeck U. Physiotherapy applied to the horse: a review. Equine Veterinary Journal. 2006; 574-580

4. Clark B. Physical Rehabilitation in Small Animal Orthopedic Patients. Veterinary Medicine. 2001; 234-246

5. Douglas S, Pippi N, Costa F, Vescovi L, Conti L, Weiss A. 2008. Therapeutic ultrasound at 1 MHz, at a dose of 0.5 W cm-2, on the bone tissue of canes assessed by optical densitometry in radiographic images. Ciència Rural. 38: 8. 2225-2231

6. DownerA. Physical TherapyforAnimals. Selected Techniques. Springfield. 1978

7. Haltrech H. Physical Therapy. Can Vet J. 2000; 41:573-574.

8. Jaegger G, Marcellin-Little DJ. Reliability of Goniometry in Labrador Retrievers. AJVR. 2002; 63 (7): 979-986

9. Jaffery N. Spinal Cord Injury in Small Animals. Current and Future Options

for Therapy. The Veterinary Record. 1999; 14:183-190

10. Marsolais G, Devorak G. Effects of postoperative rehabilitation on limb function after cranial cruciate ligament repair in dogs. JAVMA. 2002; 9: 1325-1330

11. Mills L, Taylor M. Rehabilitation and Physical Therapy. Veterinary Clinics of North Am Small Animal Practice. 2005.

12. Prentice WE. 1990. Sports Medicine. Therapeutic Techniques. St Louis Missouri. Ed Mosby. 1990

13. Ruiz Perez M. Canine Rehabilitation and Physical Therapy. Ed Intermedica. 2011

14. Santoscoy Mejia EC. In: Orthopedics, Neurology and Rehabilitation in Small Species. Dogs and Cats. 2008

15. Server R, Hury L. 2016. Introduction to the World of Rehabilitation - Physiotherapy. Asis Training.

16. Speciale J, Fingeroth J. Use of Physiatry as the Sole Treatment Three Paretic or Paralyzed Dogs with Chronic Compressive Conditions of the Cauda Portion of the Cervical Spinal Cord. JAVMA. 2000; 217:221

CLINICAL AND HEMATOLOGIC EXAMINATION IN CANINES AND FELINES UNDER PHYSIOTHERAPY TREATMENT

Marchionni, M[1],[2]

[1]Chair of Small Animal Clinic; [2]Physiotherapy Laboratory- LAFIVET, Faculty of Veterinary Sciences, National University of La Plata.

SUMMARY

Whatever the reason for an owner's consultation, both clinical veterinarians and specialists must know how to recognize the general physical condition of the patient and the pathologies that may be accompanying the condition for which they were consulted. This is of vital importance since the final result of any therapy will depend on many factors, among them the general condition of the patient, the intervening pathologies, the character of the animal and the predisposition of the owner (3 and 7). It would be useless to expect good results in our therapy if we do not first manage to stabilize underlying diseases such as anemia or infection, just as examples. For this reason, in the following chapter we will try to refresh the most important parameters to consider, both in the general clinical evaluation and in the laboratory studies. These will be useful to know in a global way the initial state and during the therapy of our patient, and thus to be able to understand the possible reasons of his evolution and if necessary to redirect the therapy. It is advisable to be methodical at the time of each evaluation in order not to overlook hidden or incipient pathologies and from their detection to be able to adapt the therapeutic plan to a more appropriate one for our patient. This is the only way to obtain the best final result. In this chapter we will offer the basic useful tools for the evaluation, follow-up and general monitoring of the patient in rehabilitation.

INTRODUCTION

Physical therapy and rehabilitation are aimed at recovering or improving the physical and clinical condition of the animals (10 and 12). When planning the protocol to be implemented, it is necessary to consider the patient as a whole, evaluating the condition in question together with the clinical history. The performance of a complete clinical examination, with the use of the complementary methods required for each case, together with an orthopedic and neurological examination, will provide a broader view of the particular situation of each animal. This comprehensive approach will allow to adapt the therapeutic plan to each patient. This avoids aggravating the clinical picture during therapy due to possible concomitant diseases and detecting the appearance of new ones during the course of therapy, which favors better results for the patient in question. At the time of establishing the therapeutic plan, a series of objectives should be established according to each case (5). These should be reevaluated and, if necessary, reformulated according to the evolution and response of each patient. It is important to know and inform the owner of the expectations of each treatment, since the goals of therapy may range from simple recovery of basic functions (being able to feed, urinate or defecate by their own means) to recovery of full functions prior to the current disease. As previously mentioned, the evaluation of the patient for physical therapy will be comprised of a general clinical examination (8), to be developed in this chapter; and a specific one (orthopedic and neurological).

Clinical Evaluation:

A medical history will be made for each patient in which data from the first day of consultation and the modifications detected in each new evaluation throughout the treatment will be recorded. Knowing the complete clinical history, the definitive diagnosis, the patient's limitations and the owner's expectations will allow to plan a rehabilitation plan according to the patient's needs, which can be modified according to the changes that are generated. Although during the clinical examination all the present conditions will be evaluated (4), in this chapter we will mention those that can influence directly or indirectly on the results of the therapy.

-Owner's data.

-Resena: Consider the species, breed, sex, color, date of birth, name, reproductive status, origin of the animal, character. The same management will not be performed in canines and felines, since the acceptance of the treatment will be different. In canines of brachycephalic breeds, the occurrence of Brachycephalic Syndrome will be evaluated. In medium and large breeds, mainly Great Dane, Doberman, German Shepherd, etc., Dilated Cardiomyopathy will be considered, or in small breeds, Mitral Valvulopathy. In Persian felines or their crosses, if suspected, polycystic renal disease will be ruled out by ultrasound, since it will condition the medical management. Regarding sex and reproductive status, consider in entire males prostatic diseases, mainly carcinoma that can generate vertebral metastasis with a consequent neurological affection. The age of the patient will help to evaluate the degree of demand to be exercised, mainly in the use of the treadmill or pool. The character of the animal is no less important, since indocile or aggressive patients will limit the use of this therapy. Lastly, it is important to consider the animal's origin in order to evaluate certain endemic diseases such as Leishmaniasis, Ehrlichiasis, Leptospirosis, Dirofilariasis, Dioctophimosis and Hepatozoonosis, among others.

Anamnesis: The knowledge of the data corresponding to the condition for which the therapy will be carried out (current anamnesis) as well as the history of previous diseases and treatments (pre-medical anamnesis) are equally relevant. Preliminary: To inquire about the health status, history of organic and neurological pathologies, surgical history, chronic diseases and to know the treatment given, the response and evolution. The occurrence of concomitant diseases in some cases will help to opt preferentially for a particular therapy. Current: We will deepen in the data related to the pathology in question. Onset, course, type of treatment (medical-surgical), response to treatment, existence of previous episodes. In many cases, inquiring about the environment in which the animal lives and its performance in it will allow a better understanding of the response to treatment and evolution. Examples are patients who may live in homes with stairs and slippery floors, or those who have access to parks and large expanses for movement. It is important to ask if the patient has undergone behavioral changes, as this may be associated with chronic pain.

- General Physical Examination: This step should be performed consciously and methodically so as not to miss any relevant situation. It will be repeated with each visit of the patient in order to detect changes or appearance of new pathologies. The inspection is the first step and of great importance. The observation of the patient's performance, from the waiting room and in the office is very useful. It will provide data on the evolution of certain parameters such as the ability to sit up and walk, the posture in station and in motion, antialgic postures (eg. Xiphosis), coordination, claudication, muscle atrophy, edema, alterations in the alignment of the limbs, etc.. These parameters may or may not be easily recognizable on inspection, and will be reevaluated in specific orthopedic and neurological examinations. The animal's behavior may reflect the presence of pain (11), refusal or acceptance of treatment. Changes in physical constitution (body condition and muscle mass), coat condition, presence of abnormal noises (coughing, sneezing, rales) may be detected for further evaluation. After the general inspection, the animal's weight, temperature, hydration status subjectively (evaluation of mucous membranes and skin folds), attitude, respiratory rate, mucous membrane coloration, capillary refill time, heart rate and pulse will be evaluated first.

Then the examination will be performed by apparatus or topographic regions.

-Respiratory apparatus: the conformation of the nostrils and trachea will be evaluated. The frequency, amplitude and type of respiration will be recorded and the lungs will be auscultated for abnormal sounds. Respiratory disorders limit the patient's aerobic capacity by altering ventilation and thermal regulation. Therefore, it will be taken into account to limit the use of hydrotherapy and tape in case the severity of the condition warrants it. -Cardiocirculatory system: By means of auscultation of heart sounds and frequency, and pulse evaluation, murmurs, tachycardia or arrhythmias may be detected. These cardiovascular pathologies may require the limitation of the patient's physical demands. It should be considered that cardiac pathologies, such as hypertrophic cardiomyopathy in felines is often asymptomatic, so the necessary studies should be performed when suspected. The incidence of thrombosis may limit the use of certain therapies such as massotherapy. -Digestive system: examination of the mouth for dental abnormalities is important as it may alter the patient's diet. The presence of lingual or gingival ulcers may be associated with uremic syndrome due to renal failure and medications. Superficial and deep abdominal palpation should be performed to look for pain, thickened bowel loops, organomegaly or masses. In patients with neurological or orthopedic involvement of the hind limbs it is important to evaluate the colon by palpation, as they may present coprostasis and require diet or supportive treatment. It is not uncommon for megacolon to occur in older felines. Likewise, evaluation of the perineal area for entities that may cause pain or discomfort should be carried out. The presence of vomiting or gastroenteritis should be studied in more depth and as a priority. Always consider the possible relationship with medications prescribed during therapy (2). -Urogenital apparatus: palpation of the bladder projection area is of great importance, mainly in patients with hind limb paresis or paralysis. These will require special maneuvers for bladder emptying or remain in certain situations with urinary catheters. In this regard, it is a priority to train the owner to avoid excessive retention of urine with the consequent occurrence of bacterial contamination. Feline owners should be evaluated and asked about the possible occurrence of changes in urination, such as pollakiuria, dysuria or hematuria, which could often be triggered by stress. In the case of female dogs in particular, both mammary rows will be palpated for early detection of neoplasms that may be of higher priority than the original picture. The vulva is also checked for pathologic discharges. In male canines, mainly those older than 2 years, rectal examination is important to detect prostatomegaly, cysts, abscesses or tumors that can generate some type of neurological compression or aggravate coprostasia. Testicular palpation is equally important. -Skin: it will be oriented to the detection of ectoparasites that could transmit diseases that alter the rehabilitation process such as Hepatozoonosis, Ehrlichiosis (9) and Hemobartonellosis in felines. The presence of dermatitis, open or infected wounds should also be evaluated for their treatment since they will impede the application of most therapies. The revision against the grain of the skin, mainly in depigmented areas and mucous membranes will allow detecting the existence of hematomas, petechiae or ecchymosis as a consequence of coagulopathies. Its importance lies in the fact that the presence of such alterations makes it impossible to perform local therapies, mainly massotherapy, in addition to requiring urgent diagnosis and treatment. -Lymph nodes: palpation of the superficial lymph nodes (mandibular, prescapular, axillary, popliteal and superficial inguinal) in search of lymphadenomegaly, pain, alteration in shape or consistency may reflect inflammatory, infectious or tumor processes. As for the neurological and orthopedic evaluation, they will be performed in a complete way and will be the ones that will provide more data in relation to the patient's own evolution in physical therapy and rehabilitation.

Hematologic and laboratory evaluation:

There are certain important parameters to consider during rehabilitation therapy that will serve as monitoring. In turn, each patient will require specific evaluations based on his or her particular condition. In this section the most relevant laboratory values and their modifications associated exclusively to the conditions that bind us will be analyzed in a general way. For more detailed information it will be necessary to consult specific bibliography (1 and 14).

- Hemogram:

Hematocrit: Before its interpretation it is necessary to know the physiological variations. In canines the reference ranges oscillate between 37 and 55%. An exception are Greyhounds and Whippet whose values range between 48 and 66%. In the case of puppies up to 6 weeks it is normal to find values between 24 and 34 % without clinical significance. The reference values in felines vary between 30% and 45%. The main causes of increase will be dehydration, fear, excitement, shock, intense exercise, absolute polycythemia, use of anabolic steroids and altitude. It is important to consider that the blood sample should be taken prior to the patient's physical exercise to avoid falsely increased values as a consequence of splenocontraction and hemoconcentration. Of greater relevance and frequency is the finding of low hematocrit values that may be due to anemia, pregnant patients, effect of tranquilizers and anesthetics or hemolysis due to an erroneous sample collection. The most frequent entity to be found will be anemia. From a more detailed study of the blood, considering the hemoglobin values (canines from 12 to 18 and felines from 8 to 15 g/l), the hemacytometric indices and the reticulocyte count, the type of anemia (hypochromic or normochromic; macrocytic, microcytic or normocitic; and regenerative or arregenerative) will be differentiated to deepen the study and determine the cause of its origin. The most frequent are hemolytic due to hemoparasites and immune-mediated, chronic hemorrhages in cases of long-standing wounds or medicated ulcers, or arregenerative due to renal failure, endocrinopathies or chronic diseases. It is important to evaluate in case of felines with hemolytic anemia due to Mycoplasma Hemofelis the possible concurrence of Feline Immunodeficiency Virus (FIV) and/or Feline Leukemia Virus (FLEV).

Platelets: Normal platelet count is in the range of 200,000 to 500,000 platelets/ul in canines and 300,000 to 700,000 in felines. Counts below 50,000 to 20,000 generate spontaneous bleeding. The presence of thrombocytopenia warrants a more detailed study of the cause and its resolution, but in the first instance it will require cessation of training and rest to avoid the risk of further bleeding.

Leukocyte count: The normal leukocyte count in canines is 6000 to 17000/ul, being that of greyhounds slightly lower (3,500 to 10,500/ul). In felines it ranges from 5,500 to 19,500/ul. An increase in the normal count, called leukocytosis, is usually due to an increase in the number of neutrophils and/or lymphocytes. The main causes of leukocytosis related to the rehabilitation patient are bacterial infections, steroid effects and tissue necrosis or severe inflammation. It is important to consider that intense exercise may temporarily increase the total leukocyte count without exceeding the normal range. Other causes can be myeloproliferative disorders such as leukemias, feline infectious peritonitis, and feline hyperthyroidism. In canines and felines under prolonged stress, due to illness or chronic pain, the release of endogenous steroids may be reflected in the appearance of a stress leukogram (leukocytosis with neutrophilia, lymphopenia, eosinopenia and monocytosis). On the other hand, leukopenias can be associated with severe bacterial infections and anaphylaxis, leukemias, viral diseases (FIV, VILEF, feline panleukopenia, canine distemper, canine parvovirus and canine infectious hepatitis). For a correct interpretation of the picture, the differential leukocyte count should be evaluated (13).

- Blood biochemistry:

Urea: it is synthesized in the liver from the metabolism of ammonium coming from endogenous proteins or from the diet. Reference values range from 15 to 40 mg/dl in canines and 30 to 65 mg/dl in felines. Its increase may be due to pre-renal causes such as high protein or low carbohydrate diets, intestinal bleeding (by use of NSAIDs or glucocorticoids), fever and tissue necrosis, prolonged exercise, use of catabolic drugs (glucocorticoids), hypoadrenocorticism and dehydration. A common disease in elderly felines is hyperthyroidism, which often goes unnoticed by the owners. Renal causes, chronic renal insufficiency can be decompensated by the use of drugs. Post renal causes can be urethral obstructions by compression in prostatic neoplasms or in urinary infections with lithiasis, frequent in patients with neurological deficiencies and in felines. Bladder rupture should be considered in patients with recurrent infections and manual bladder emptying or with obstructive processes. Reductions in urea levels may reflect the intake of a diet poor in protein, the use of anabolic steroids that divert protein intake to protein synthesis, hepatic insufficiency or portosystemic shunt among the most frequent.

Creatinine: derived from the catabolism of muscle creatine. Its reference values range from 0.5 to 1.5 mg/dl in canines and felines, and its variations are not subject to dietary changes. It may be increased in the presence of acute or chronic renal failure, dehydration, urinary flow obstructions, bladder rupture and intense exercise.

Total Plasma Proteins: the reference values range between 5.7 and 7.7 g/dl in canines and 5.8 and 8 g/dl in felines. It is important to consider that the values of total proteins are in the lower range in animals younger than 6 months. Albumins and the rest of the proteins (except immunoglobulins) are of hepatic synthesis, and in case of variations in the levels of total proteins it is necessary to evaluate the different fractions to determine which is the cause. In general, hyperproteinemia is caused by dehydration, inflammatory, infectious, neoplastic or autoimmune diseases and the use of anabolic steroids. On the other hand, hypoproteinemia can not only alter the patient's performance, but is also responsible for poor wound healing. The main causes are starvation, intestinal malabsorption, hepatic processes, congestive heart failure and losses through urine, intestine, burns or severe inflammatory and exudative lesions, hemorrhages or septic conditions. The values of albumin in both species range from 2.5 y 4 g/dl. The causes correlate with those mentioned in hypoproteinemia.

Glycemia: reference values are 60 to 100 mg/dl for both species. The most frequent causes of increase in patients under rehabilitation therapy are fear, excitement or stress mainly in felines, severe trauma or chronic stress, both reflecting the increase of catecholamines. The use of certain drugs such as glucocorticoids, xylazine, morphine and ketamine also cause hyperglycemia. Convulsive patients may manifest elevations in glucose levels due to the release of epinephrine. It is also important to consider that immediately after extreme exercise (mainly recorded in greyhounds) plasma glucose levels are increased. The degree of increase should also be considered y the other, much more frequent, causes, such as diabetes, obesity, etc. The causes of hypoglycemia are also varied (hyperinsulinism, hypoadrenocorticism, hepatic dysfunction, hypoglycemia in puppies, sepsis, etc.) and will reflect concomitant conditions not necessarily related to rehabilitation therapy, but of no lesser importance, which will prevent the correct evolution of the treatment.

Other determinations are of little importance to be performed routinely in our patients, unless their evaluation is useful to extend a study. Certain considerations should also be taken into account.

Bilirubin and bile acids are increased by the administration of steroids, in addition to the increase due to the most known causes such as hepatocellular disease, biliary obstruction and hemolysis. Triglycerides increase immediately post-exercise y cholesterol in trained animals is lower than average in canines y may be increased after prolonged high-dose glucocorticoid therapy.

Enzymes:

The reference values vary according to the method used for their measurement, therefore, it is advisable to consider the value indicated by each laboratory.

Aspartate Aminotransferase (AST): This enzyme is found in higher concentration in cardiac, skeletal and liver muscle, so its increases are related to damage in them. It can be found increased in the presence of skeletal muscle damage or intense exercise, accompanied by changes in Creatinine Phosphokinase (CPK). Its peak activity appears 12-24 hours after stimulation, remaining for 5 to 6 days.

Alkaline phosphatase (ALP): to evaluate its importance, the different existing isoenzymes must be considered. Hepatic and bone isoenzymes are the most important, being also present those produced by intestinal, renal and placental cells. In our canine patients, their increases are probably associated with the hepatic isoenzyme induced by corticosteroids and the bone isoenzyme. The latter appears elevated in cases of extensive or generalized bone diseases, bone tumors, in fracture healing processes, or normally in growing animals, being able to exceed up to 6 times the estimated adult value in both dogs and cats under 6 months of age. Likewise, it is commonly found increased in canines subjected to a high level of training. It should be noted that corticosteroid-induced hepatic isoenzyme is not present in felines.

Creatinine phosphokinase (CPK): It has isoenzymes in skeletal muscle, myocardium and CNS. Its increase is normal in animals younger than 6 months, and it has higher values in males than in females. For a correct evaluation of its activity, it should be taken into account that its increases occur 6 to 12 h after the triggering event and disappear after 48 h unless the stimulus remains. The most frequent causes of increases are skeletal muscle injuries, whether inflammatory, traumatic, endocrine, nutritional, etc., as well as from intense exercise. It is important to consider that hypothyroid patients tend to show higher values as their rate of catabolism is reduced, along with degenerative changes in muscle fibers. CNS disorders, mainly those that generate seizures may increase their activity from muscle damage.

Lactate Dehydrogenase (LDH): It has 5 isoenzymes in different tissues (skeletal and cardiac muscle, liver, erythrocytes, pancreas, bone and lung), but its increases mainly reflect cardiac or skeletal muscle and liver disease. It may appear falsely increased in cases of hemolysis.

From these determinations, it is important to consider that the 3 enzymes (CPK, AST and LDH) reflect skeletal muscle disorders (inflammatory, traumatic, hereditary, metabolic, endocrine, etc.). CPK is the most specific, but it is also the most labile, reaching a peak of maximum activity 6 - 12 h after the onset of damage. AST and LDH have lower peaks but last longer. In the case of AST its peak value appears at 12-24 h and persists for 5 to 6 days, and in the case of LDH it may take several days to appear, but its increase persists for several weeks.

-Electrolytes and metals:

Calcemia: although it is not a routine determination in general check-ups, some reasons why it may be increased can be associated to pathologies that concern us. The reference values in canines are 8 to 12 mg/dl and 7.2 to 12 mg/dl in felines. They can appear increased in cases of tumor bone lysis or not (such as osteomyelitis), diffuse osteoporosis, which usually occurs as a consequence of immobilization for a long time as a consequence of neurological or musculoskeletal damage. Glucocorticoid therapies can generate hypocalcemia.

Potassemia: values in canines range from 14 to 22.5 mg/dl and 14 to 21.5 mg/dl in felines. The causes of its increase and decrease are varied, but the main clinical manifestation in cats with hypokalemia is cervical ventroflexion and generalized muscle weakness, usually associated with urinary losses due to renal dysfunction or gastrointestinal losses. In the subclinical form it is more difficult to detect, but the presence of mild anemia, gradual weight loss, anorexia and reduced physical activity should be cause for investigation. It also manifests with rigid ambulatory and muscle pain (6).

Phosphatemia: its reference values are 2.5 to 5 mg/dl in dogs and 4 to 8 mg/dl in cats, doubling in the case of canines under one year of age. Its increase can be associated with the presence of osteolytic bone tumors, among other causes such as renal insufficiency, hypoparathyroidism, bladder rupture, hypervitaminosis D, etc. In felines the most common causes are diabetes mellitus, hepatic lipidosis or intake of phosphorus binders. When the deficit is marked, it manifests with myopathy, cardiac and neurological alterations and acute hemolytic anemia. Its decrease, as with calcium, can be associated with glucocorticoid treatment, rickets and osteomalacia among other causes.

-Urinalysis: It is important to consider it in our patients, mainly to monitor renal function in those under chronic treatment with NSAIDs or Glucocorticoids, as well as in patients with hind limb paralysis who are predisposed to suffer ascending urinary tract infections.

In the first case, the main attention should be paid to urine volume (presence of polyuria: > 50 ml/k/day), color (transparent, very clear amber), associated with a low urine density (< 1015) and proteinuria (due to the loss of albumen in the glomerular filtrate). In the case of urinary tract infections the urine is usually cloudy in appearance due to the presence of crystals, mucus, casts, bacteria and cells (pyocytes, red blood cells, desquamation cells). The odor is usually ammoniacal and there is proteinuria (associated with globulins, homoglobin, among others). Periodic urinalysis is suggested in these patients for early diagnosis of urinary tract infections. Considering that these patients are very prone to urinary tract infections, empirical treatment of urinary tract infections is not recommended, but rather urine culture and antibiogram (with samples obtained by bladder puncture) to avoid antibiotic resistance and to obtain a better response to treatment. The presence of myoglobinuria is frequent in rhabdomyolysis due to exertion or severe muscle trauma.

CONCLUSIONS

Structuring the work methodology allows a more orderly follow-up of the patient and a global knowledge of the patient. Having a clinical history, which is completed in a methodical way with each visit of the patient, added to the performance of routine studies in the form of monitoring and special studies in case of need, will allow a personalized therapy and therefore obtain results according to what is expected for each patient. It is true that many times time is not an ally, but it is important to be aware that the creation of a systematic evaluation routine will become in the future an incorporated way of working and really reduce the possible errors or failures in the therapy. Treating each patient based on his or her needs and particularities will bring us closer to success as a final result.

BIBLIOGRAPHY .

1.   Bush. B.M. Interpretation of laboratory tests for small animal clinicians. Spain. Harcourt Ed. 1999.

2.   Cortadellas O. Rational use of NSAIDs. Argos Rev. 2012. (N° 140). p 46-48.

3.   Drum MG. Physical rehabilitation of the canine neurologic patient. Vet Clin North Am Small Anim Pract. 2010 Jan;40(1):181-93.

4.   Ettinger, SJ. Physical exploration of the dog and cat. In Ettinger. SJ and Feldman. Treatise on Veterinary Internal Medicine. 6th ed.Ed Elsevier. 2007. p: 2-9.

5.   Levine D, Millis DL, Marcellin-Little DJ. Introduction to veterinary physical rehabilitation. Vet Clin North Am Small Anim Pract. 2005. 35(6):1247-54.

6.   Norsworthy, G.D; Crystal, M.A; Fooshee Grace, S; Tilley, L.P; The Feline Patient. Intermedical Ed. 2009.

7.   Olby N, Halling KB, Glick TR. Rehabilitation for the neurologic patient. Vet

Clin North Am Small Anim Pract. 2005. 35(6):1389-409.

8.   Rijnberk A, de Vries HM. Anamnesis and body exploration of small animals. Ed Acribia 2nd ed. 1991.

9.   Robin W, Meinkoth A. Anemias Caused by Rickettsia, Mycoplasma, and Protozoa. In Weiss D.J, Wardrop K.J. Schalms, Veterinary Hematology. 6th ed.Ed Wiley-Blackwell. 2010. P: 199-210.

10. Ruiz Perez. M. Rehabilitation and canine physiotherapy. Ed. Intermedica. 2011.

11. Salazar Nussio V, Rioja Garcia E, Taboada.FM. Pain management. Rev. Argos.2014. (N 158, volume II). p 22-29.

12. Santoscoy Mejia E. Orthopedics, neurology and rehabilitation in small species. Ed. Manual Moderno. 2008.

13. Schultze E. Interpretation of Canine Leukocyte Responses. In Weiss D.J, Wardrop KJ Schalms, Veterinary Hematology. 6th ed. Ed Wiley-Blackwell. 2010.p: 321-334.

14. Sodikoff ChH. Diagnostic and laboratory tests in small animal diseases. Mosby Ed. 1996.

CLINICAL CLINIC OF SPINAL MEDULA AFFECTIONS IN SMALL ANIMALS

De Palma Viviana Edith[1,2]

[1]Chair of Small Animal Clinic;[2] Chair of Semiology, Faculty of Veterinary Sciences, National University of La Plata.

INTRODUCTION

Spinal cord related disorders are one of the most frequent reasons for consultation in daily clinical practice and are a challenge for the clinician. It is essential to perform a very complete and detailed clinical/neurological examination and anamnesis in order to establish the specific site of the lesion and thus indicate the appropriate complementary study, thus allowing to reach the definitive diagnosis. Often due to lack of financial resources it is difficult to obtain a definitive diagnosis.

Spinal cord disorders are a reason for consultation both in emergencies and in clinical practice, due to the diversity of entities that affect it.

Abstract

The summary is of great importance, since almost all of its parts have great relevance, because it is precisely the beginning of a diagnostic path that can lead to errors.

Species:

Spinal cord problems are very different if our patient is a canine or a feline. From the breeds with particular characteristics to the differences in the biomechanics of one species and the other, which make different processes and evolutions.

Regarding accidents in canines there is a greater probability of automobile accidents, while in felines it is more frequent to find bites caused by dogs.

<u>Race:</u>

There are many breeds par excellence at the top of the list: dushund, basset hound, as chondrodystrophic breeds, and with a tendency to be chondrodystrophic as the cocker spaniel, when a patient with a spine condition is presented for consultation the first thing we think of is a problem in an intervertebral disc. Other breeds such as French bulldogs, English bulldogs are predisposed to congenital or hereditary conditions such as block vertebrae, hemivertebrae, transitional vertebrae, etc. And some with specific entities of some breeds as for example the hereditary ataxia of Jack Russell and Fox terrier smooth hair. In felines the sacro coccygeal dysgenesis of the Manx cat. Reason why we must know the breeds with which we work.

<u>Age:</u>

The age together with other data of the resena what kind of problem we can have. In puppies congenital or hereditary conditions. Spondylosis deformans is an entity that we expect in elderly patients, as well as the presence of neoplasms.

In middle-aged animals the spectrum of possibilities is much wider. In these cases we must be careful not only with the data of the resin but also with the anamnesis and the physical examination that we will refer to later.

<u>Sex:</u>

In spinal cord problems there is no correlation with sex.

<u>Use y environment:</u>

This is an important point since a working animai does not have the same prognosis as a pet animai.

On the other hand, related to the environment in which the patient lives, an animal that lives in the street or is allowed to do so by its owners is much more exposed. Traumas caused by automobiles and fights between animals are a very common reason for consultation in the daily clinic.

<u>Coat:</u>

The color of the coat in canines of breeds such as Maltese, West Highland White terrier, Bichon frise and their crosses becomes relevant because of tremors (white shaking dog syndrome).

Anamnesis

In this part of the consultation we should try to collect as much information as possible, in some cases it will be very rich if the owners cooperate and are responsible and in other cases with polytraumatized animals that are collected from the street we have almost no data.

In problems concerning the spine, the evolution is one of the data that we cannot forget, if it is necessary to ask several times, because the prognosis of our patient will depend largely on the time elapsed since the onset of the signs or in the case of a trauma at the time of the accident. It is of interest not only the current anamnesis (the reason for consultation) but also the previous anamnesis (the history of that patient), among other data to know vaccination plan to rule out infectious diseases, medications received and their response, and any other data of importance to the case.

Physical examination

Whenever a patient who is presumed to be neurological comes for consultation, a complete general examination is necessary to allow us to correctly define the problem or to discover a second problem. In many cases the reason for consultation may not be well defined or may be accompanied by systemic signs that may be related to the condition being sought.

<u>Gait and movements:</u>

If the animal does not ambulate, that is to say that it presents incoordination in the station (astasia), we may receive it in decubitus. It is necessary to preserve the patient by placing him on a hard surface, either a stretcher or some rigid material to replace it. If the animal wanders we observe the gait, at this point we must be careful and be able to differentiate whether we are facing an orthopedic or neurological problem, for this we see the animal walk as far as possible on hard and soft floor, if you have incoordination in walking (ataxia) or absence of proprioception is a problem of neurological origin, if claudic is an orthopedic problem.

We can observe other compatible signs such as tremors, tics, chorea, etc. that can be the reason for consultation or be a finding of the consultation accompanied or not by the signs mentioned above.

Once we determine that the problem is of neurological origin the main objective is to locate the site of injury. To do this we must know the anatomy of the nervous system and then perform a series of specific semiotic maneuvers.

<u>Examination of proprioception</u>: the first test to be performed is the test of the back of the foot or hand, the latter being very simple to perform and also very reliable, ideally repeat the maneuver two to three times to be sure to observe the response. There are other tests (postural reactions) that depending on the size of the patient some are more accessible than others. They are: jump test, hemimarcha-hemisthesia, optical and/or tactile extensor force and wheelbarrow test.

<u>Inspection and palpation of the spine:</u>

Before palpation, we must always inspect the spino in detail to observe any change in shape (kyphosis, scoliosis, lordosis) or any deformation after trauma that is evident.

Palpation is performed vertebra by vertebra by palpation-pressure in search of sensitivity, palpating spinous and transverse processes. We can perform a combined palpation maneuver by placing a passive hand on the ventral region to be explored and an active hand that exerts pressure on each vertebra. We must reproduce the passive movements of the joints for this we stop in the cervical region (C1-C5) and perform flexion and extension and laterality of the neck. To evaluate the dorsolumbar spine, we perform the flexion and extension movements taking the four limbs, with the animal in lateral decubitus.

<u>Examination of reflexes</u>:

The semiological value of reflex examination is to locate the exact site of the lesion. In the thoracic limb we perform the withdrawal or flexor reflex with which we evaluate the plexus in its totality, the other reflexes both the bicipital and the tricipital are not always constant and thus commit errors. In the pelvic limb the reflexes that we evaluate are the patellar, sciatic, anterior tibial and flexor or withdrawal. Then we evaluate the anal reflexes waiting for the contraction of the anal sphincter and indicating its absence in S1-S3. It is important at this point to return to the anamnesis and ask if the animal, to the owner's knowledge, has conscious urination and/or defecation, asking pertinent questions. In felines in many cases they refuse to be restrained, which is why we can perform the reflexes with the animal in suspension.

The reflexes have a numerical value according to the observed response:

0: absence of reflex (areflexia)

1: decreased response (hyporeflexia)

2: NORMAL (euroreflexia)

3: exaggerated response (hyperreflexia)

4: clones.

Another very useful reflex is the panniculus or cutaneous muscle reflex, which, like the reflexes mentioned above, contributes to finding the lesion.

<u>Sensitivity evaluation</u>:

This is perhaps the most complex moment of the consultation and the one to which we must pay the utmost attention in order not to make mistakes, because the prognosis of our patient depends on this examination.

We evaluate the superficial sensitivity with relatively light stimuli (needle prick) paying attention to the animal's reaction. To perform the test of deep painful sensitivity we need a hemostatic forceps and we must make pressure on both the medial and lateral portions of the limb to explore and carefully observe the attitude of the animal that may react biting, crying, or manifesting in some way the pain, if the animal responds only by withdrawing the limb, the response will be negative, ie our patient presents only withdrawal or flexor reflex and does not feel pain. As mentioned above this test indicates prognosis and the decision to intervene or not surgically.

Most frequent spinal cord diseases in the clinical practice

For a better approach we divide the spine into regions: Cervical Region (C1-C5), Cervicothoracic Region (C6-T2), Thoracolumbar Region (T3-L3), Lumbar Region (L4-S3) and Coccygeal Region (cauda equina).

Cervical region (C1-C5):

In lesions located in this region the patient may present at consultation with ataxia, neck stiffness with manifestation of pain on palpation, proprioceptive deficit in the four limbs, tetraparesis, hemiparesis, with signs of NMS, there may be muscle atrophy due to lack of use if we are facing a certain time of evolution. Reflexes are normal or increased. Tetraplegia is rarely observed because the lesion was so severe that it affected the respiratory center. Loss of voluntary control of defecation and urination may occur.

Cervicothoracic region (C6-T2):

If the lesion is located in this portion (cervical intumescence), the brachial plexus is located in this region, where the IMN (lower motor neurons) are located. The lesion located in this segment causes ataxia, tetraparesis with signs of NMI in the thoracic limbs, i.e. normal or diminished reflexes and signs of NMS in the pelvic limbs, with normal or exaggerated reflexes. Also as mentioned in C1-C5 may appear loss of voluntary control of both defecation and urination.

Thoracolumbar region (T3-L3):

This is the region par excellence where automobile accidents or other traumas such as bites occur in felines in particular. The clinical signs that we find are paraparesis or paraplegia with normal or exaggerated reflexes (NMS), muscular atrophy depending on the evolution. The anal reflex is normal or exaggerated as well as the bladder.

In this region a phenomenon called "Shiff-Sherrington" occurs, in these cases the animal presents signs of NMS in pelvic limbs and rigidity and hyperextension of the thoracic limbs. This is due to the lack of inhibition by the lesion of the ascending lumbosacral fibers that synapse with the IMN of the extensor muscles of the thoracic limbs. However, after about 24 to 48 hours, a compensatory mechanism takes place that returns the relaxation of the thoracic limbs to normal.

Lumbar region (L4-S3):

Patients with lesions in this spinal segment present paraparesis or paraplegia with signs of NMI, i.e. with normal or depressed reflexes in their pelvic limbs and normal thoracic limbs. Accompanied by bladder and anal dysfunction. Muscle atrophy may be present according to the time elapsed. Conscious proprioception is usually absent or reduced.

Coccygeal region (cauda equina):

In this region are the lumbosacral and coccygeal nerve roots in the final portion of the spinal cord. Lesions in this segment are very similar to those mentioned for the previous segment. Ataxia, paraparesis, bladder and anal dysfunction and tail paralysis can be observed.

Congenital and/or hereditary malformations:

The most frequent vertebral bone anomalies are: hemivertebrae, block vertebrae, transitional vertebrae, butterfly vertebrae, variation in the number of vertebrae, lack of fusion of sacral vertebrae, spina bifida, etc. The most affected breeds are French bulldog, English bulldog, pug pug, Boston terrier.

Atlantoaxial subluxation: agenesis or hypoplasia of the odontoid apophyses, causing instability and / or spinal cord compression. The most affected breeds include: yorkshire terrier, poodle toy and other small breeds. In felines, sacrococcygeal dysgenesis of the mans cat presenting deformed sacrum, absence of tail or caudal vertebrae and spina bifida may appear.

Trauma:

It is very common in daily clinical practice to find a polytraumatized patient, after primary care as mentioned above, and once the patient is stabilized, we proceed to the approach of the spine according to the type of injury, the spinal cord involvement and clinical symptomology. In many cases they are animals that live in the street and we do not have any anamnestic data, even in some cases when the accident occurred. In felines, canine bites are more common.

Intervertebral disc disease:

This is one of the most frequent pathologies that can cause disc herniation or extrusion causing spinal cord compression. It is classified in two types I and II. Type I occurs in chondrodystrophic breeds: Duschund, Beagle, Pekingese, I hasa apso, Shih tzu. and breeds that tend to be chondrodystrophic such as Cocker Spaniel and toy poodle. Type I discopathy is characterized by degeneration with rupture of the dorsal annulus fibrosus and extrusion of the nucleus pulposus into the spinal canal. Type II disc disease occurs in large breeds, is characterized by fibroid degeneration, there is a bulging of the intervertebral disc without complete rupture of the annulus fibrosus.

Both types are of acute onset or when the animal manifests clinical signs it does it from one moment to another and it is when the owner brings it for consultation. The signs vary according to the involvement of the medullary canal and the time of evolution.

Infectious diseases:

There are several infectious agents that can cause disturbances in the spinal cord. In some cases they present with neurological signs only and in other cases they are accompanied by systemic signs that may point to an infectious disease. From laboratory findings such as the presence of thrombocytopenia to the visualization of ectoparasites that help us to arrive at an accurate diagnosis.

Parasites: *toxoplasma gondii, neospora cainum.*

Bacterial: *Staphylococcus albus, S aureus, S epidermidis, Pasteurella, Actinomyces, Nocardia, Brucella canis.*

Fungal agents that appear are : *Cryptococcus neoformans, Histoplasma capsulatum, Blastomyces dermatidis.*

Viral agents: *canine distemper, feline immunodeficiency.*

Other agents isolated very frequently in our environment are *Hepatozoon canis and Erlhichia canis.*

Vascular diseases:

One of the most common entities to observe in the office is the fibrocartilaginous embolism that occurs in both canines and felines, is an acute condition, the neurological signology is variable according to the damage produced. In canines it usually occurs in large to giant breeds. The neurological deficit is in most cases symmetrical or bilateral, although it can occur asymmetrically, unilaterally.

Other vascular causes are hemorrhage or infarction and vascular malformations.

Degenerative:

Spondylosis deformans is one of the most common pathologies in elderly patients. Osteophytes are formed affecting the intervertebral spaces, forming osseous bridges in many cases, causing intense pain. Preferably large to giant breeds are affected, although they can occur in other breeds and sometimes they are radiographic findings.

Neoplasms:

Neoplasms are suspected in geriatric animals and large breeds, although we cannot rule them out in younger animals and small breeds, not in all cases we have access to complex complementary studies to arrive at a definitive diagnosis.

Among the most frequent neoplasms we find extramedullary neoplasms that compress the canal from the outside and intramedullary neoplasms that compress from the inside. The extramedullary ones: osteosarcoma, lymphoma, lymphosarcoma, fibrosarcoma, hemangiosarcoma. The intramedullary ones: meningiomas topping the list, neuroepitheliomas, astrocytomas, ependymomas and metastases of lymphomas or hemangiosarcomas.

In these patients we must observe for clinical signs accompanying the neurological signs and evaluate when neoplasia is suspected whether it is primary or metastatic.

Alimentary: hypervitaminosis A in felines whose diet is only liver, present confluent exostoses that manifest themselves most frequently in the cervical and thoracic spine, this disease can progress to other regions of the spine, ribs and other joints.

Diagnostic planning

Regarding the arrival of the diagnosis with the resena, a well detailed anamnesis, it is fundamental as mentioned in previous paragraphs, the basic clinical examination followed by a detailed neurological examination guides us to the site of injury and to decide the appropriate complementary method since some of them are not entirely accessible because they are expensive or not accessible in our environment. Among the most used complementary methods, imaging is of great help: X-rays define many pathologies such as traumatic lesions, vertebral instabilities, spondylitis disc, dysplasias, etc.

Other more sophisticated imaging studies such as computed axial tomography (CT) or magnetic resonance imaging (MRI) are very useful to arrive at an infinite number of diagnoses, but in our environment we do not always have them.

The laboratory is of great help in the infectious processes both hemogram and blood biochemistry as well as serology and PCR study.

BIBLIOGRAPHY .

1. Cote, Etienne. The Consultant in the Veterinary Clinic. Editorial intermedica Bs As, Argentina. 2010.

2. Couto G, Nelson R.: Medicina interna de pequenos animales. Editorial intermedica , Bs As, Argentina. 2002.

3. Ettinger SJ, Feldman EC. A treatise on veterinary internal medicine. Diseases of the dog and cat. Elsevier. Madrid. Sixth edition. 2007.

4. Miller P E, Tilley L P, Smith, F W K. The veterinary consultation in 5 minutes canine and feline. Editorial intermedica Bs As Argentina. 2007.

5. Rijnberk A, de Vries, H W Anamnesis and body exploration of small animals. Editorial Acribia, S.A Zaragoza (Spain). 1997.

6. Ruiz Perez M. Canine Rehabilitation and Physical Therapy. Ed Intermedica. 2011

7. Santoscoy Mejia EC. In: Orthopedics, Neurology and Rehabilitation in Small Species. Dogs and Cats. 2008

8. Kenneth J. Drobatz, Merilee F. Costello Emergencias em medicina felina.

Ed Intermedica. 2012.

PAIN CONTROL IN CANINES AND FELINES

Montiel ME[1,2]

[1]Chair of Small Animal Clinic; [2]Physiotherapy Laboratory - LAFIVET, Faculty of Veterinary Sciences, National University of La Plata.

SUMMARY

The interpretation, assessment and management of acute and chronic pain is one of the challenges facing veterinarians today. By performing a good interpretation, based on the clinical examination and pain assessment scales, the acting physician can orient himself to establish an adequate and individualized therapy for each patient. In this way, multimodal therapies can be used with the aim of reducing the dose, toxicity and side effects in the short and long term use of these drugs. Currently, multiple tools are being used in combination with medical therapy, such as physiotherapy, hydrotherapy, acupuncture, which together will help to improve the well-being of these patients.

INTRODUCTION

The World Health Organization (WHO) defines PAIN as "an unpleasant sensory and emotional experience, associated with a real or potential tissue injury", we can also add that it is a subjective sensation and different in each individual.

Pain is very difficult to measure and value as it will depend on many independent issues such as race, age, socio-environmental status, chronicity of pain, and even the emotional state of the patient (stress, overcrowding).

The management of pain and its early interpretation provides numerous benefits such as reducing hospitalization times, avoiding prolonged anorexia (especially in felines) and even avoiding self-mutilation or aggression on the part of the patient.

Definitions and classifications

According to the duration they can be divided into

- Acute pain: It is the effect of a traumatic, surgical or infectious event that begins abruptly. It is relatively brief and almost always disappears with anesthetics (4).

- Chronic pain: Pain that persists beyond the usual time of acute illness, or beyond a reasonable time for an injury to heal, or is associated with a chronic pathologic process. It is rarely permanently relieved by analgesics, but may respond to a combination of analgesics, tranquilizers, psychotropic drugs, physical therapy and environmental manipulation (4).

According to their pathogenesis

- Neuropathic: pain caused or initiated by dysfunction or primary lesion in the central and peripheral nervous system. Example: amputations, plexus avulsion, diabetic neuropathy.

- Nociceptive: results from direct activation of pain receptors (nociceptors) in the skin or other tissues in response to tissue injury, usually accompanied by inflammation. It can also be called *infammatory pain.*

According to your location

- Somatic: Somatic pain is pain originating in the skin, muscles, joints, ligaments or bones.

- Visceral: the pain originates in an internal organ or serous membranes (e.g. pancreatitis, peritonitis, hepatitis).

Evaluation

The evaluation of pain in small animals is difficult, even more so in feline patients. In veterinary medicine, several types of objective and subjective pain assessment scales are used. The most commonly used scales are the Melbourne (Table 1), University of Colorado (Table 2) and Glasgow scales.

Whichever scale is used, it should be as representative as possible, should fit in with the work methodology of the staff in charge of the patient and above all should facilitate the observation and interpretation of the response to the analgesia administered. Any of the tables chosen should be selected differentiating between canines and felines, the ideal is to always use the same one to be familiar with its use. Pain assessment is dynamic and should be performed as often as necessary, for example before and after certain medication, and during hospitalization or treatment.

Pardmeters of pain in canines

-   Elevated resting heart rate

-   Tachypnea or shallow breathing

-   Altered capillary filling time

-   mydriasis

-   Hypertension

-   Constipation

-   immobility, reluctance to move

-   Unlimited movements of body areas, claudication, claudication

-   Alteration of character, aggressiveness or dependence on the owner

-   Anorexia

-   Vocalizations

-   Abnormal postures, praying position, syphosis

-   Automutilacion

-   Appearance of ventricular extrasystoles

Feline pain parameters

-   High heart rate

-   Tachypnea and shallow breathing

- Altered capillary filling time

- Mydriasis

- Curvature of the spine

- Moving out of the way or hiding

- Depression

- Stop grooming

- Elbows backward

- Head hung, in position lower than its body

- Vocalization

- Excessive licking of an area (chronic pain)

- Aggressiveness

These parameters are more evident and easier to identify in acute pain, in chronic pain the signs are more obvious or "silent" (hyporexia, apathy, exercise intolerance or refusal to play certain games).

Table 1. Melbourne pain rating scale (5,6).

Category	Feature	ValoraciPn
Physiological data	a. Normality	O
	b. Dilated pupils	2
	c.* Increased heart rate over basai	
	> 20%	1
	> 50%	2
	> 100%	3
	d.' Increased respiratory rate over basai	
	> 20%	1
	> 50%	2
	> 100%	3
	e. Elevated rectal temperature	1
	f. SalivaciPn	2
Response to palpation		
	Does not change	O
	Reacts[1] when touched	2
	React before touching	3
Activity* Activity		
	Resting, sleeping, semi-conscious.	O
	Resting, awake	1
	Eating	O

	Restless	2
	Turning. rolling around	3
State of mind		
	Submissive	O
	Very friendly	1
	Frightened	2
	Aggressive	3
Posture		
	a. Protect the affected area	2
	b.* Lateral decubitus	O
	External debit	1
	Lying downZSittingZIn station head upright	1
	At the station, crestfallen	2
	Moving	1
	Abnormal posture (praying, bowing)	2
Vocalization*, "	Does not vocalize	O
	Vocalizes when touched	2
	Intermittent vocalization	2
	Continuous vocalization	3

*Choose one option; ** Barking alert is excluded

1 Licking, scratching, rubbing, nibbling, straining, protects itself.

Minimum Value = O; Maximum Value = 27

Table 2. Feline acute pain scale. University of Colorado (5)

Scale (0-4)		Psychological and behavioral attitude	Response to palpation	Body tension
0		O Contento y quiet if not heeded o Interested in or curious about the environment.	O The wound or any other place can be palpated without any concern for it	**Minima**
0-1		o subtle signs, difficult to detect in hospitalized patients o isolation from the environment or change in normal routine, detectable in the home o in the hospital may be quiet or slightly unstable. o less interest in the environment, but looks around to see what's	o may or may not react to palpation of a wound	**Slight**

		going on.		
1-2		o decreased response, prefers to be alone o motionless, loss of brightness in the eyes, eyes partially or almost closed, balled up position or retracts all four legs under the body, shoulders hunched, head slightly lower than the rest of the body, tail curled tightly.	o Responds aggressively or tries to escape if the painful area is touched or near the painful area o Tolerates attention, even improves its mood when stroked as long as the area is avoided.	**Mild to moderate** Re-evaluate analgesia plan
		around the body o rough or partially lifted hair o insistently licking the painful or irritated area o anorexia or low interest in food	painful	
2-3		o Maùlla, grune or bufa Constantly when unattended o May bite the wound, but is quiet when left alone	o Grune or grunt to non-painful palpation, (may have allodynia, hyperalgesia, or fear of worsening pain. o React aggressively to avoid palpation, y run away to avoid any contact,	**Moderate** Re-evaluate the analgesic plan,
3-4	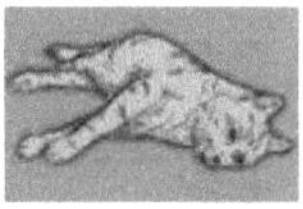	o prostrate o May be unresponsive or oblivious to surroundings, difficulty distracting from pain	o May be unresponsive to palpation o There may be stiffness to avoid	**Moderate to severe** Stiffness may be present to avoid painful

		o more receptive to attention (cats with bad temperament or even feral cats allow themselves to be touched)	painful movements.	movement Reevaluate analgesic plan

Medication management

Medical management should be oriented to "multimodal or multifunctional" therapies (Figure 1).

Specific therapy to treat the underlying cause should alleviate the discomfort, and in some cases analgesics, in addition to other adjuvant medications and even other therapies, may be required for effective pain management.

Figure 1. WHO Pain Analgesic Scale (8,10)

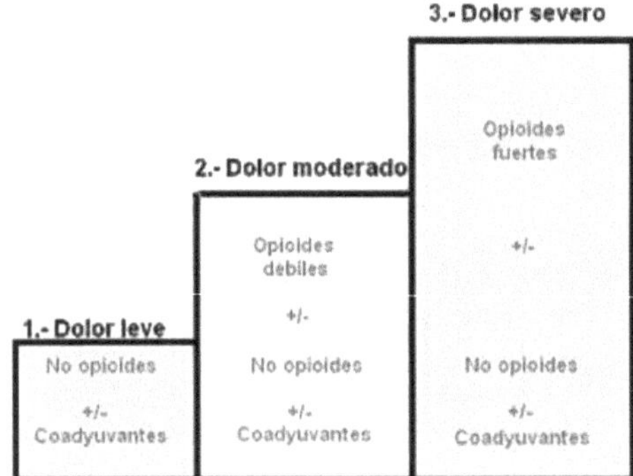

Opioids

Opioid receptors are found in the CNS at both the spinal and supraspinal levels, and are also present in peripheral tissues.

In animals there are 3 types of receptors responsible for pain transmission and perception: μ (0P3), κ (0P2) and α(0P1). The analgesic actions are mainly given by: μ (OP3), κ (OPP2), but also most of the unwanted effects.

Indications: these drugs promote effective analgesia for moderate to severe pain, this efficacy depends on the dose, route of administration and interval. They are used in peri-operative management, as sole therapy or in multimodal analgesia protocols.

Side effects and contraindications: the most common effects are vomiting, nausea, gasping, bradycardia and dysphoria, these signs are closely associated with the doses and routes of administration. They are CNS depressant drugs and therefore their main effect is respiratory depression as they desensitize the respiratory bulb to co2 levels, generating from a decrease in respiratory frequency to apneas.

The μ-agonists would be contraindicated in pacreatitis since they increase the contractility of the sphincter of Oddie, which increases the possibility of retrograde flow towards the pancreas (8). This does not occur with meperidine or nalbuphine.

With morphine, hypotension may occur due to histamine release during rapid intravenous injection, and in the case of morphine syrup, dysphoria may occur at high doses. Meperidine should NOT be used by intravenous administration.

In long-term use, especially in patients with chronic pain, opioids can cause constipation.

As for butorphanol and nalbuphine, they have a greater sedative effect and cause only mild analgesia.

Table 3. Opioid doses, routes and intervals in canines and felines (2.7)

	Dose	Via	Interval
Morphine	0.1-1 mg/kg can	SC, IM, POJV	4-6 hs
	0.1 - 0.2 mg/kg fel	SC. IM	6-8 hs
Meperidine	3.5- 10mg/kg	IM	2-3 hs
Fentanyl	2-10 mcg/kg can	IV	20-30 min
	1-5 mcg/kg fel	IV	20-30 min

Remifentanil	0.25	IV	Infusion
	0.1 mcg/kg/min		continues
Butorphanol	0,2	IM,SC,IV,PO	2-4 hs
	0.6 mg/kg can	IM,SC,IV	2-4 hs
	0.2-0.8 mg/kg fel		
Nalbuphine	0.5-1mg/kg can	IM,SC,IV	4-6 hs
	0.5-3 mg/kg fel	IM,SC,IV	2-4 hs
Tramadol	1-2mg/kg	IM, SC, IV,PO	6-8 hs
D-propoxyphene	0.04-1mg/kg can	IM,SC,IV	4-6 hs
	2 mg/kg fel	IM, SC, IV	4-6 hs

Alpha-2 Agonists

They are sedatives, muscle relaxants and generate excellent visceral analgesia. Its disadvantage is its short duration as analgesic and its main undesired effect is an initial hypertension, followed by a moderate hypotension. It can also cause a decrease in respiratory rate, which associated with other depressants could cause apneas. Xylazine and medetomidine produce emesis in canines and even more in felines. They induce hyperglycemia and hypoinsulinemia.

Table 4. Doses. Routes and intervals of alpha 2 Agonists in canines and felines (2,7)

	Dose	Via
Xylazine	0.4-1 mg/kg	IM, EV (lower doses)
Medetomidine	10-40 mcg/kg can	IM
	40-80 mcg/kg fel	IM
Dexmedetomidine	5-20 mcg/kg can	IM

NSAIDs (non-steroidal anti-inflammatory drugs):

They are analgesics, anti-inflammatory and antipyretic drugs that produce their effect through the inhibition of the Cyclooxygenase enzyme (C0X1 C0X2). Cyclooxygenases are enzymes involved in the formation of prostaglandins, thromboxanes and leukotrienes from arachidonic acid. Cyclooxygenase C0X1 is the one most associated with the protective effects of the organism, such as renal and gastrointestinal protection (housekeeping functions) or involving cellular homeostasis. In contrast, cyclooxygenase C0X2 is more associated with inflammatory effects.

Canines and mainly felines are deficient in glucuronyl transferase, the enzyme responsible for the glucuronide conjugation of some drugs to non-toxic metabolites. For this reason paracetamol (fig.2) should not be used in felines, since when only a fraction of the drug is conjugated, the rest will try to be metabolized in the liver by cytochrome P450, where the NAPQUI component (N acetyl-benzoquinonimine) starts to accumulate, causing damage to the hepatocyte, methemoglobinemia (more marked in felines) and cell death.

Figure 2. Degradation pathways of paracetamol (acetaminophen)

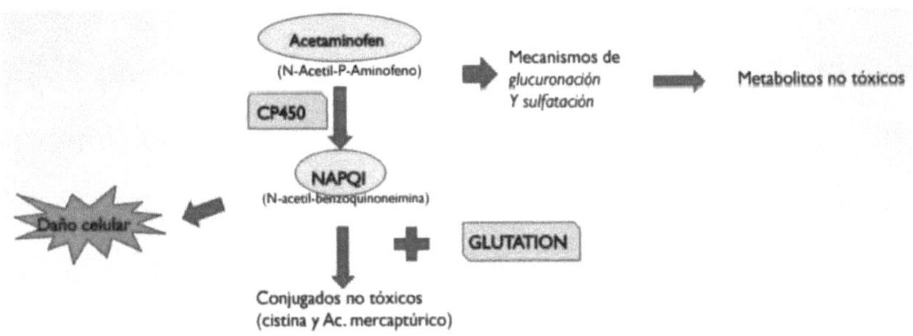

Indications: used in the management of peri-operative pain, oncology patients, chronic pain and patients with pain and associated inflammation.

Contraindications and side effects: The most common effects are gastric irritation and alteration, gastroduodenal ulcerations and renal insufficiency. It should not be used in patients with hypovolemia and hypotension, nor with pre-existing renal or gastrointestinal pathologies. They can also present a decrease in platelet aggregation (aspirin, ketoprofen) and therefore can induce post-surgical hemorrhages.

Drug	Indication	Species, dose and route	Frequency
Ketoprofen	Chirurgic and chronic pain	2 mg/kg IV,SC,IM fel-can 1 mg/kg oral	S.I.D., max. 3 days
Meloxicam	Chirurgical, acute pain	0.1-0.2 mg/kg IM, SC, IV, PO can fel 0.3 mg/kg SC	S.I.D Single dose
	Chronic pain	0.2 mg/kg PO can 0.1 mg/kg PO 0.1 mg/kg PO fel O.05 mg/kg PO	Attack dose S.I.D w/tracking Attack dose S.I.D w/tracking
Mavacoxib	Chronic pain	2 mg/kg PO can	Day 0, then day 14, then one per month x 5 more doses.
Carprofen	Chirurgical Pain	4-4.4 mg/kg SC IV PO can 2.2.2 mg/kg SC,IV,PO can 2-4 mg/kg SC, IV fel	S.I.D., max. 4 days B.I.D max. 4 days Single dose
	Chronic pain	4-4.4 mg/kg PO can 2-2.2 mg/kg PO can	S.I.D, seek min effective dose B.I.D, effective min. dose
Etodolac	Chronic pain	10-15 mg/kg SC, PO can	S.I.D
Fibrocoxib	Chirurgical Pain	5 mg/kg PO can	S.I.D. max. 3 days
	Chronic pain	5 mg/kg PO can	S.I.D, effective min. dose
Meglumine	Pyrexia	0.25 mg/kg Sc can-fel	Single dose

from			
Flumixin	Chirurgical pain, ophthalmologic pain	0.25-1 mg/kg SC can- fel	S.I.D or B.I.D, a or two doses
Ketorolac	Surgical pain	0.3 mg/kg IV, IM, SC can 0.2 mg/kg IM fel	B.I.D, two doses
	Panosteitis	5 mg/can of 20-30 kg 10 mg/can + 30 kg	S.I.D max. 3 days
Pyroxicam	Inflammation in lower urinary tract	0.3 mg/kg PO can	S.I.D, for 2 days, then c/ 48 hs.
Paracetamol	Surgical pain, acute and chronic pain	10-15 mg/kg PO can. NOT IN FELINOS	c/8-12 hs
Aspirin	Surgical pain acute and chronic pain	10 mg/kg PO canines.	c/12hs

Adjuvant Drugs

They are defined as drugs whose main indications are other than pain, but which exert analgesic actions in certain painful disorders. They are drugs that can be included in the analgesia protocol, which by themselves are not considered analgesic, but in conjunction with opioids, aines, among others. (5)

• Dissociative anesthetic: It is an NMDA receptor antagonist, and at very low doses can make a substantial contribution to analgesia by minimizing CNS sensitivity. They are used for chemical restraint, anesthetics and analgesics. Its major effect is amnesia, analgesia and catalepsy. Their undesirable effect is increased heart rate, increased blood pressure and sometimes apneas (in rapid IV bolus).

Indications: as part of perioperative multimodal therapy, in trauma, or as part of the desensitization protocol in patients with chronic pain.

Ketamine: Used at low doses does not produce catalepsy, nor unconsciousness. Dose: 0.1-1 mg/kgc4-6 h IM, EV, PO.

1-4 mg/kg EV or 5-10 mg/kg PO in felines, interval according to need and in acute pain.

•   Antoconvulsants: (gabapentinoids) act through the modulation of GABA receptors, reducing glutamate, noradrenaline, substance P and calcitonin inhibition, thus electrochemically modifying the conduction pathways, modulating pain integration, providing analgesia, anxiolytic and anticonvulsant effects.

Indications: used in the multimodal protocol for pain of neuropathic origin (diabetic neuropathy, amputations, pelvic trauma), arthritic pain (gabapentin) and oncological pain.

Advantages: they can be combined with almost all other analgesics, without antagonism, helping to reduce the minimum effective doses of both groups of drugs.

Gabapentin: Dose: 3-10 mg/kg c/ 8-12 o24 h canines

5 mg/kg c/12 hs felines

Pregabalin: Dose 4 mg/kg c 12h canines

2 mg/kg c 12 h felines

- Antiviral: (Amantadine), is an antiviral that was developed to inhibit the multiplication of influenza A virus in humans. It is effective in the treatment of drug-induced extrapyramidal effects and in the treatment of Parkinson's disease. It exerts its analgesic effect by antagonism of NMDA receptors in a manner analogous to ketamine (7).

Amantadine: 3-5 mg/kg c 24h canines

2- 5 mgkgc 24h in felines x 21 days

Environmental Management

For acute pain, in the hospitalized patient, it is suggested to try to seek the greatest possible comfort. For this, the resting place should be clean and dry, should have a padded place to lie down, try to respect light schedules and be isolated from loud sounds. In the case of felines, in addition, have clean and dry sanitary stones (if he can get up and go to them), place a box and a rolled up cloth as a nest, to allow him to curl up.

In cases of chronic pain, especially for felines, the aim is to achieve a "facilitated environment", especially in cases of osteoarticular affection, avoiding high places or placing steps, so that they have good access to food and to their litter tray.

This type of environmental modification improves the patient's quality of life and avoids repeated stressful situations.

CONCLUSIONS

Animals, like humans, feel pain. Pain is associated with important disadvantages such as delayed recovery from diseases, scarring, decreased immunity, increased predisposition to infections, alterations of the ventilatory pattern, constant hyperglycemia, among other alterations. Acute pain has a protective, biological function, but when this pain is perpetuated over time, it becomes a pathological function that brings with it many alterations in the life of the individual and his interaction with the environment.

Patients with chronic pain also suffer from depression, anxiety, self-mutilation, aggressiveness and many other behavioral alterations that go hand in hand with a detriment to their quality of life. To generate an effective pain relief, it is necessary to have the basic knowledge for its interpretation, assessment and quantification; and then, to establish its possible treatment. The combination of several drugs is more effective because when acting by different mechanisms they have a sum of beneficial effects for the patient.

Treatment must be individualized and selected according to the damage received, the time elapsed and the response of our patient.

But pain management should not only be focused on the use of medications and their environmental management, but also on supplementing pain with other therapies, such as surgery, physiotherapy, hydrotherapy, massage and acupuncture.

BIBLIOGRAPHY .

1.   Cambridge, AJ; Tobias, KM; Newberry, RC; Sarkar, DK; Subjective and objective measurements of postoperative pain in cats. j am vet surg Assoc 2000; 217:685-690.

2.   Karol A. Mathews. Pain management in cats. Pain: Evaluation and treatment in small animals - 1° ed. - Buenos Aires: Intermedica, 2004 p.149-159.

3.   Aigé V, Cruz I: "Pain in small animals: neuroanatomical basis, recognition and treatment. Consulta Difus. Vet 9 (78); 2001. P. 63-70

4.   Hellyer PW, Robertson S, Fails AD, Lamont LA, Mathews KA, Skarda RT. Pathophysiology, pharmacology, and treatment of pain. Manual de Anestesia y analgesia en pequenas especies. 1st ed. Espanol, editorial El Manual Moderno Mexico, 2013. P. 82-131.

5.   Hellyer P.Objective, categoric methods for assessing pain and analgesia. In Gaynor JS, Muir WW, eds. Handbook of Veterinary Pain Management, St. Louis Mo: Mosby; 2002. p. 82-107.

6.   Hansen BD.Assessment of Pain in Dogs: Veterinary Clinical Studies. ILAR JOURNAL 44; 2003. P. 197-205.

7.   Lascelles DX, Gaynor JS. Chronic Pain Management. Manual de Anestesia y analgesia en pequenas especies. 1st ed. Espanol, editorial El Manual Moderno Mexico, 2013. P. 147-157.

8.   Mathews k, Kronen P, Duncan Lascelles, Nolan A, Robertson S, Steagall P, Wright B, Kazuto Y. Guidelines for recognition, assessment and treatment of pain. Journal of Small Animal Practice, Vol. 55, 2014 p. E14- E34.

9.   Otero PE. Analgesic Drugs. Pain: Evaluation and treatment in small animals - 1° ed. - Buenos Aires: Intermedica, 2004 p. 93- 102.

10. Puebla Diaz F. Pain, Types of pain and therapeutic scale of the W.H.O. Iatrogenic pain. Oncologia, 2005; 28 (3):139-143.

INTRAOCULAR PRESSURE IN SPORTS

Cassagne P[1,2]

[1]Comparative Ophthalmology Service,[2] Servicio de Clinica e Internacion de Pequenos Animales Hospital Escuela, Facultad de Ciencias Veterinarias, Universidad Nacional de La Plata, Argentina.

RESÛMEN

Intraocular pressure (IOP) is the result of a balance between aqueous humor production and its different drainage mechanisms. Its values can vary significantly by any modification of the factors involved in its production or drainage. In glaucoma, IOP values increase. Taking into account the hydrostatic forces involved in the formation of aqueous humor, IOP is considered to be partially dependent on systemic blood pressure. In canines undergoing physical training it has been shown that the pressure exerted with the use of a collar on the neck region transiently increases IOP values.

It is considered that the cause of this is related to increases or decreases in venous pressure which would generate parallel variations in IOP. In both canines and felines it has been shown that body position has a significant effect on IOP and higher values were found in dorsal decubitus.

Studies have been performed in humans during isometric exercise and IOP values have increased along with systemic blood pressure values. On the contrary, IOP values decrease during dynamic exercise. In canines and felines subjected to physical training, no measurements have been made on how IOP values vary.

INTRODUCTION

Aqueous humor is a translucent fluid with a density somewhat greater than that of water, secreted by the ciliary processes found in the posterior chamber, which flows through the pupil to the anterior chamber where it is

"

drained by the corneoscleral trabecular meshwork and the uveoscleral vias. Its main functions include its participation in the nutrition and removal of debris from avascular structures such as the cornea and lens, its optical properties and the maintenance of its own transparency. There are small variations between the osmolarity of aqueous humor and that of plasma, which allows us to associate that the rate of solute transfer from plasma is related to the rate of aqueous humor production (3).

Aqueous humor could be considered as the product of plasma ultrafiltration (3). This activity is mediated by the blood-aqueous barrier formed by the two layers of ciliary epithelium surrounding the ciliary processes and their tight intercellular junctions. This barrier reduces the concentration of proteins in the aqueous humor from the plasma which, because of their size, cannot pass through it and prevents the passage of many substances into the aqueous humor (13). There are situations that can modify the composition of the aqueous humor, such as variations in the metabolic needs of the cornea or lens, certain pathologies such as diabetes mellitus that lead to hyperglycemia or alterations in the speed of blood supply to the ciliary processes (3). In several species the plasma concentrations of many macromolecules such as proteins, lipids, glucides, amino acids, among others, exceed the levels in the aqueous humor (3). Most of these different concentrations are due to the action of the blood-aqueous barrier, such is the case of proteins, immunoglobulins, enzymes and urea. Glucose enters by diffusion and plasma concentrations would be expected to be similar, but are found in lower concentrations due to metabolic consumption by the cornea and lens. Amino acids are found in a higher concentration than plasma concentrations, associated with a recruitment-type activity by the vitreous humor. Ion concentrations are variable, with sodium representing the highest concentration compared to cations such as potassium, calcium and magnesium. The anions are chloride, bicarbonate, phosphate, ascorbate, and lactate. Glucose metabolism in different intraocular structures generates a higher level of lactate in the aqueous humor than in the plasma. Ascorbate through an active transport process has higher levels than plasma (3). Various concentrations of ions and macromolecules were found in canines. IOP is maintained at a relatively constant level because the rate of aqueous humor formation is in equilibrium with its drainage (3,14).

Physiology of aqueous humor

The production and drainage of aqueous humor involves both various anatomical components of the anterior ocular segment and numerous endogenous compounds.

Aqueous humor formation occurs in the ciliary body and basically involves three mechanisms, one active (selective transport against a concentration gradient) and two passive (diffusion and ultrafiltration). The transport against a gradient of some plasma molecules such as sodium ion is one of the main mechanisms involved in the active formation of aqueous humor. The Na-K activated ATPase located in the non-pigmented ciliary epithelium is the main responsible (3). Among the passive mechanisms, diffusion allows isosoluble molecules to cross the cell membrane of the ciliary epithelium in favor of a concentration gradient. Ultrafiltration allows the passage of macromolecules and water in response to a hydrostatic force.

The enzyme carbonic anhydrase catalyzes the formation of carbonic acid from carbon dioxide and water. By dissociating carbonic acid into negatively charged bicarbonate it allows these ions to pass into the aqueous humor and positively charged sodium ions and water to follow the bicarbonate into the posterior chamber (13,14).

Hydrostatic forces against an oncotic pressure gradient together with active transport cause aqueous humor to flow through the ciliary body into the posterior chamber.

Aqueous humor production is regulated by humoral factors and by the autonomic nervous system.

Aqueous humor leaves the eye by different pathways. The conventional drainage pathway involves the passage of aqueous humor from the posterior chamber through the pupil to the anterior chamber. The aqueous humor circulates through the anterior chamber by a process known as thermal circulation, which is generated by the temperature difference between the cornea, which is in contact with atmospheric air, and the temperature of the iris. This generates an upward flow of aqueous humor near the iris and a downward flow of aqueous humor near the cornea. The aqueous humor leaves the anterior chamber through the ciliary fissure of the trabecular meshwork and, after a series of interconnected vessels, leaves the eye through the scleral venous plexus, which flows into the systemic venous circulation. More than 80% of the aqueous humor is drained by this route in canines. In felines most of the aqueous humor flow (> 97%) is produced by this route (11). Conventional drainage pathways could also be termed pressure dependent, since they are dependent on IOP values. The sites of main resistance to aqueous humor outflow are the scleral venous plexus, the endothelium of the angular aqueous plexus and the extracellular matrix.

In canines, approximately 20% of the remaining 20% uses the uveoscleral or non-conventional pathway, which crosses the interstices of the ciliary muscle and is distributed to the supraciliary or suprachoroidal space (3). On the contrary, in felines this pathway represents a very small percentage of the flow of aqueous humor (<3%) (11). This drainage pathway is independent of IOP values.

Intraocular pressure (IOP)

We can define intraocular pressure (IOP) as the result of a balance between the production of aqueous humor and its different drainage mechanisms. Any alteration in any of these factors may result in variations in IOP (17).

In canines and felines, normal IOP values vary depending on the type of tonometer selected for its measurement. In canines the values range from 10 to 20 mmHg and in felines between 12.3 ± 4.0 mmHg, using the Tonopen applanation tonometer, however there are certain variations in IOP that may have different origins. One of them is related to the circadian cycle, and in canines as in humans, the values are somewhat higher during the morning and decrease during the night (3). On the contrary, in felines, higher values were found at night and a gradual decrease in IOP occurred during the day (11). In general, age also significantly varies aqueous humor production and IOP values are lower in canines and elderly felines. Variations in IOP according to body position have been found in both canines and felines, obtaining higher values in dorsal decubitus than in the rest of the positions studied (ventral and lateral decubitus) (2). Certain anesthetic drugs can decrease IOP. Intraocular inflammation (uveitis) reduces IOP. Severe alterations affecting ocular blood flow will reduce IOP(14).

In glaucoma, both the production and drainage of aqueous humor are altered, and the acute elevation of IOP in canines is manifested with buftalmo, variable mydriasis, corneal edema and episcleral congestion among other signs (14). In felines, glaucoma occurs less frequently and the initial clinical signs are less evident than in canines, with little corneal edema and less episcleral congestion (11). In both species, signs of chronicity such as butthalmos and exposure keratopathy in uncontrolled glaucoma are inevitable. Increases in IOP can affect visual capacity in a variable way, due to retinal and optic nerve involvement by two mechanisms. Firstly, ischemic changes in the retina are considered as a result of compression due to increased IOP, which initially alter the axoplasmic flow of the axons and the irrigation of the optic nerve head, although it can also compromise the choroidal irrigation and produce ischemia in the photoreceptors. The pathophysiology of this process is complex, in general it is considered that the loss of vision is usually a consequence of degeneration and death of ganglion cells. At the molecular level, when these cells die, they release glutamate and excitotoxins that bind to receptors, one of the most important of which is N-methyl D-aspartate (NMDA), which is highly permeable to calcium, which enters the cell generating enzyme activation and the inevitable apoptosis. This process is self-perpetuating and generates apoptosis in adjacent ganglion cells that were not affected. Even if IOP values are normalized under medical treatment, this mechanism may be responsible for irreversible vision loss (10, 14, 16). Felines are more resistant in terms of susceptibility to glaucomatous damage.

with relative preservation of retinal ganglion cells after acute elevations of IOP, on the contrary, in chronic uncontrolled glaucoma the progressive loss of vision is almost irreversible (11).

INTRAOCULAR PRESSURE IN SPORT.

IOP may vary due to any modification in some of the factors involved in its production or drainage. Generally speaking, it is much more frequent that the drainage is decreased than that the production mechanisms are increased (15).

It has been shown that pressure exerted with the use of a collar over the neck region in canines transiently increases IOP values. This is because compression of the jugular vein may result in vascular congestion and increased choroidal blood volume. This mechanism would produce a rapid increase in IOP. Another proposed mechanism is that as venous pressure increases, the episcleral veins present increased resistance to aqueous humor drainage, although this latter mechanism generates IOP elevations more slowly (15). The importance of venous pressure is that its increase or decrease modifies IOP values in parallel (18).

Considering a patient undergoing physical training, the use of a sling would be recommended to avoid compression and possible repercussions on IOP.

On the other hand, dorsal decubitus position in both canines and felines manifested increased IOP values. The exact mechanism of action by which body position may alter IOP is unclear. However, previous studies in humans have suggested that IOP readings were altered as a result of body position through one or more of the following mechanisms: changes in episcleral venous pressure, choroidal vascular volume, gravity or body fluid shift, increased intraperitoneal pressure, and in turn, central venous pressure (2).

Variations in blood CO_2 concentration variably modify central venous pressure (CVP) and IOP. Increases in ETCO2 (end-expiratory partial pressure of CO2) increase IOP values, which occur as a result of choroidal vasodilatation or increases in CVP (18).

Taking into account the hydrostatic forces involved in the formation of aqueous humor, IOP is considered to be only partially dependent on systemic arterial pressure (3). Severe decreases in blood pressure values that do not allow adequate tissue perfusion may decrease aqueous humor production and IOP values. In these severe cases of hypovolemia, ETCO2 decreases and PVC decreases as well (14,18).

In humans, several studies have been conducted on how physical training varies IOP and its relationship to blood pressure. They have found significant differences between the effects of isometric exercise and those of dynamic exercise in relation to IOP variations.

In isometric exercise, IOP increases in parallel with increases in systemic blood pressure, both systolic and diastolic. Studies have shown that IOP keeps increasing continuously while physical training persists and in parallel to blood pressure increases. After isometric exercise, IOP decreases, attributing to exercise a post-exercise lowering effect (1).

This IOP increase has been related to increases in choroidal sympathetic activity or myogenic autoregulation that minimize choroidal congestion so that IOP increases are minimal (6, 7).

In contrast, during dynamic exercise, IOP decreases while systemic blood pressure increases (18). Considering that these studies apply to short-term physical activity, it has been shown that IOP values normalize after approximately 30 minutes (17).

Even though IOP decreases, ocular perfusion increases (9, 18) to maintain visual function during exercise. That is, despite the increase in ocular perfusion pressure during exercise, the retina maintains a constant flow. This autoregulation triggered by exercise has two mechanisms, on the one hand a vasoconstrictor effect and on the other an increase in sympathetic tone. This mechanism attempts to compensate for the decrease in the diameter of the choroidal arteries and branches of the ophthalmic artery resulting from the distribution of blood flow to the main muscles involved in exercise (4, 5, 12,18).

CONCLUSIONS

Considering that aqueous humor is the result of plasma ultrafiltration, and that IOP is maintained within normal parameters due to a controlled balance between the mechanisms of production and drainage of aqueous humor, the study of the different variants that may imply modifications in its values is of medical interest.

Taking into account the hydrostatic forces involved in the formation of aqueous humor, IOP is considered to be only partially dependent on systemic blood pressure.

In general, pathologies that generate an increase in IOP such as glaucoma are relatively frequent in daily clinical practice and usually lead to an unfailing loss of visual function, it is important to evaluate the IOP of all patients prior to undergoing any variety of physical training.

Considering that it has been proven that IOP increases transiently as a consequence of the compression generated by the use of a collar on the neck of a patient, and in particular when subjected to certain types of exercise, it is important to avoid its use in patients who could suffer from glaucoma. In this case, the use of a headgear would be indicated.

In humans, dynamic exercise has been shown to stimulate IOP decreases, which is interesting considering the possibility of subjecting a canine with this pathology to dynamic exercise (18). However, there are currently no studies carried out in canines and felines.

In relation to studies in people during isometric exercise, although it was observed that IOP increases, other trials have obtained contradictory results. Some studies have found that IOP values during isometric exercise decrease and other studies have found no significant changes in IOP. These variations are closely related to the type of instrumentation used to measure IOP. For example, with the applanation tonometer it has been found that there is a minimal decrease in IOP when the tonometer contacts the corneal surface and exerts pressure that would result in stimulation of aqueous humor drainage. While differences in the methods used in the production of isometric exercise, muscle mass involved and type of posture adopted may explain such variations in the results, the data on this subject are inconsistent (1).

Considering that the results of studies in humans on how physical training modifies IOP in both isometric and dynamic exercise are varied, and that currently no such studies have been performed in canines and felines, it would be of interest in veterinary medicine to consider the possibility of evaluating IOP values during physical training.

BIBLIOGRAPHY .

1.   Amy M. Pauli, Ellison Bentley Kathryn A. Diehl, Paul E. Miller. Effects of the Application of Neck Pressure by a Collar or Harness on Intraocular Pressure in Dogs. J Am Anim Hosp Assoc 2006; 42:207-211.

2.   Caryn E. Plummer, Alain Regnier, and Kirk N. Gelatt. The Canine Glaucomas .In: Edited by Kirk N.Gelatt, Brian C. Gilger, Thomas J. Kern , Veterinary Ofhthalmology Fifth Edition ; Two volume set , USA: Wiley-Blackwell Publishing House; 2013 .p.1050-1145207.

3.   Charles L. Martin.Glaucoma. In: Edited by Charles L. Martin, Ophthalmic Disease in Veterinary Medicine. London UK: Manson Publishing Ltd; 2010.p. 337-368.

4.   Espen F. Bakke, Jonny Hisdal, and Svein O. Semb. Intraocular Pressure Increases in Parallel with Systemic Pressure during Isometric Exercise. Journal IOVS 2009; 50 (2): 760-764.

5.   Gillian J McLellan and Paul E Miller .Feline Glaucoma - A Comprehensive Review. Vet Ophthalmol. 2011 ; 14(1): 15-29.

6.   Glenwood G. Gum and Edward O. MacKay. Physiology of the Eye. In: Edited by Kirk N.Gelatt, Brian C. Gilger, Thomas J. Kern , Veterinary Ofhthalmology Fifth Edition ; Two volume set , USA: Wiley- Blackwell Publishing House; 2013 .p.171-207.

7.   Harris, A.; Arend, O.; Bohnke, K.; Kroepfl, E.; Danis, R. and Martin, B. Retinal blood flow during dynamic exercise. Graefe's Arch Clin Exp Ophthalmol ; 1996. 234:440-444.

8.   Ivia Carmem Talieri, Cristiane dos Santos Honsho, Newton Nunes, Almir Pereira de Souza, Juan Carlos Duque. Comportamento da pressão intraocular segundo os efeitos cardiorrespiratórios e hemodinâmicos induzidos Pelaanestesia com desflurano, em cães submetidos à hipovolemia experimental. Arq Bras Oftalmol. 2005; 68(4):521-6.

9.   Kergoat, H. and Lovanski, J. V. Response of parapapillary retinal vessels to exercise. Optometryand vision science72; 1995.(4):249-257.

10. Kiel JW. Choroidal myogenic autoregulation and intraocular pressure. Exp Eye Res. 1994;58:529-543.

11. Kiel JW. Modulation of choroidal autoregulation in the rabbit. Exp Eye Res. 1999;69:413-429.

12. Kroll MM, Miller PE, Rodan I. Intraocular pressure measurements obtained as part of a comprehensive geriatric health examination from cats seven years of age or older. Journal of the American Veterinary Medical Association. 2001 ; 219(10):1406-1410.

13. Lovasik, J. V.; Kergoat, H.; Riva, C. E.; Petrig, B. L. and Geisar, M. Choroidal blood flow during exercise-induced changes in the ocular perfusion pressure. Investigative Ophthalmology and Visual Science; 2003.44 (5):2126-2132.

14. Masoud Selk Ghaffari and Ahoora Arman Gherekhloo. Effect of body position on intraocular pressure in clinically normal cats. Journal of Feline Medicine and Surgery ; 2017 : 1-3.

15. Michelson, G.; Groh, M. and Grundler, A. Regulation of ocular blood flow during increases of arterial blood pressure. British Journal of Ophthalmology; 1994. 78(6):461-465.

16. Miller PE. Structure and function of the eye In: David J. Maggs, Paul E. Miller, Ron Ofri, Editors. Maggs, Paul E. Miller, Ron Ofri, Editors, Slatter, Fundamentos de oftalmologia veterinaria. 4° edition, Barcelona (Spain): Editorial Elsevier; 2009. P.1-19

17. MillerPE. Glaucoma. In: Davidj. Maggs, Paul E. Miller, Ron Ofri, Editors, Slatter, Fundamentos de oftalmologia veterinaria. 40th edition, Barcelona (Spain): Editorial Elsevier; 2009. p 235-262.

18. Price, E. L.; Gray, L. S.; Humpries, L.; Zweig, C. and Button, N. F. Effect of exercise on intraocular pressure and pulsatile ocular blood flow in a young normal population. Optometry and vision science.2003; 80(6):460-466.

PHYSIOTHERAPY AND BODY WEIGHT ASSESSMENT IN DOGS AND CATS

Risso A[1,2]

[1] Faculty of Veterinary Sciences' National University of La Piata;[2] IGEVET- CONICET

SUMMARY

Nutrition is a key point for overall animal welfare and the basis for preventing common nutritional diseases in the daily veterinary practice. Thus, overweight and excessive accumulation of adipose tissue are related to a wide variety of pathologies, including musculoskeletal disorders. Physiotherapy together with a balanced diet are the most effective treatment for these conditions. At the time of consultation it is necessary to control the weight and keep a periodic record of it to ensure the success of the treatment. This chapter describes on the one hand' the relationship between weight and musculoskeletal disorders in companion animals and on the other hand' the different methods used to measure body condition in both dogs and cats.

INTRODUCTION AND DEVELOPMENT

Within the veterinary Physiotherapy and Rehabilitation as non-invasive and complementary disciplines to the medical clinic' and whose main objectives are prevention' diagnosis and treatment of pathologies that cause pain or functional loss to improve the quality of life of patients, control and weight loss are key points in the treatment.

Any patient susceptible to a musculoskeletal, neurological or functional injury can be treated with physiotherapy. The physiotherapeutic treatment never follows a pre-established protocol but is a constant, continuous and individualized process that evolves according to the patient's progress.

Weight control in animals is not only important for aesthetic reasons, but it directly affects the quality of life and the good functioning of the organism. Maintaining an ideal weight for each particular animal will allow us to increase both physical and cardiovascular resistance directly impacting on the daily activity of the individual, the recovery of joint mobility and gait reeducation among others.

Frequently, both clinicians and owners relate overweight with various pathologies such as cardiac, respiratory and even reproductive problems. However, the importance of weight control for the proper functioning of the musculoskeletal system is often overlooked. It has been mentioned that every kilo of excess weight an animal has translates into joint overstrain and increased wear of the cartilage. This leads, in the first place, to a notable decrease in muscular activity and mobility. Secondly, it contributes to atrophy due to disuse and reduces the support capacity of the musculature. And finally, it leads to joint instability and degenerative diseases such as osteoarthritis.

The maintenance of weight within the limits is always advisable and when suffering from a joint pathology such as osteoarthritis of the knee in canines, it becomes a necessity because overweight leads to joint overload that accelerates the degenerative process (2,10). In addition, the excess in the diet of energy and nutrients such as calcium together with an accelerated growth due to excess in the diet are risk factors in the development of the disease.

The role of the owners is fundamental to control food consumption together with a rigorous management of diets and rations by the veterinarian (4).

There are many and varied risk factors associated with overweight and obesity, the most relevant being overeating, endocrine diseases such as hypothyroidism and diabetes mellitus, lack of physical exercise, ethological and genetic alterations, sex of the animal, reproductive status and breed, among others (1,4,5).

Different authors point out that weight control influences the joint disease process by reducing stress on the joint. Currently, in dogs with osteoarthritis of the knee it has been proven that weight reduction reduces symptoms and increases quality of life and life expectancy. On the other hand, within the dietary prescriptions it is very important to maintain a high level of proteins, minerals and vitamins for the maintenance of the musculature and the general health of the animal.

In cats, excessive body weight is also associated with increased stress on the joints. As a consequence, it is common for overweight cats to claudicate. In turn, older cats are more likely to suffer from osteoarthritis (12).

Lower mobility reduces the animal's energy requirements and, therefore, if the daily food portions are not adjusted to this condition, the animal will gain weight, which in turn generates lower mobility (12).

Therefore, it is necessary to monitor weight control in dogs and cats and different methods are currently used. The physical and biochemical methods measure the body condition of the animal and serve to evaluate the degree of fatness of the same (11).

Among the most widely used and accessible physical methods are, on the one hand, the measurement of body weight, which should not be considered as an effective method because there is no ideal weight standard given the individual body differences of each animal and, on the other hand, the morphological scales. The latter are based on external morphological characteristics of the individuals to analyze and score them, the scales can be up to 5 points or up to 9 points. The ideal weight animals will be at 3 points and 5 points for each scale respectively. It is at these points (3 or 5) that ribs are easily palpable with a slight fat cover, as well as bony prominences, but are not visible on inspection. The presence of subcutaneous fat is palpable but not in large amounts and the hourglass shape is evident. Animals scoring between 4-5 on a 5-point scale, or 7-9 on a 9-point scale, are considered overweight or obese (8).

The use of photographs allows to observe the condition of an animal and is a useful and easily accessible tool. Although palpation is recommended for a better estimation of the level of body fat, the observation of photographs collaborates with the estimation of the body condition in a first instance, in order to make a correct and more accurate evaluation in the office (3).

Figures N^0 1 and 2: Mestizo female, 6 years old, estimated body condition 6

At this point the thoracic cage and the spine are palpated slightly because of the fat deposits that accumulate in the area. The waist of the animal is hardly observable.

Figures № 3 and 4: Female Labrador Retriever 9 years old, body condition estimated 7

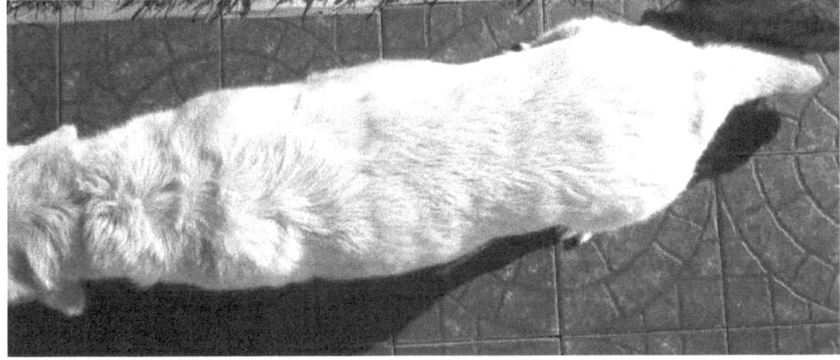

At this point the thoracic cage and the spine are difficult to palpate due to the fat deposits that accumulate in the area. The waist of the animal is barely visible or not observed.

Figures N^0 5 and 6: 7 year old female mestizo, estimated body condition 3

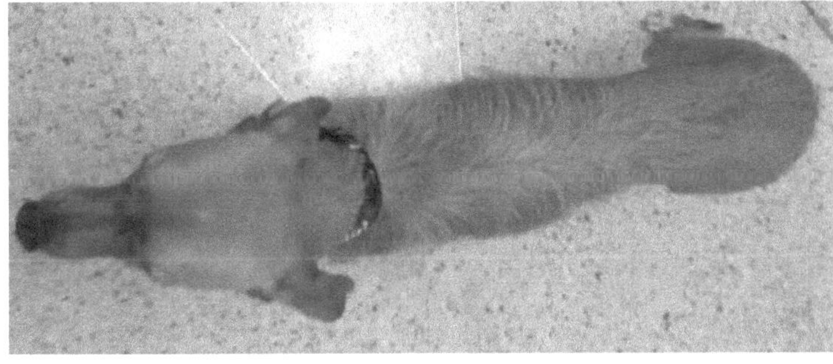

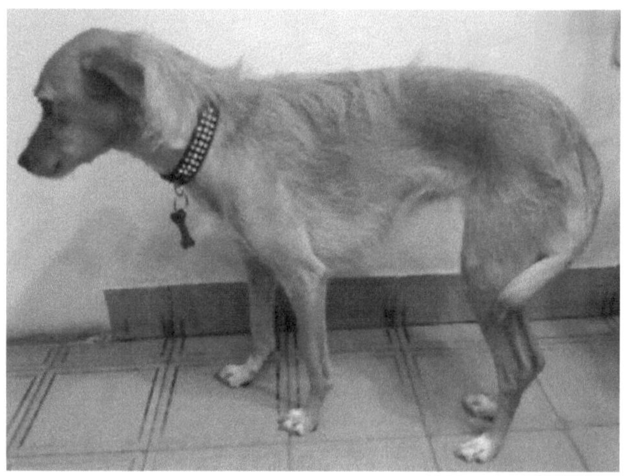

At this point the thoracic cage, spine, scapulae and pelvis are easily observed in short-haired animals. Loss of muscle mass and absence of palpable fat in the thoracic area can also be observed.

As for biochemical methods, they are based on obesity-related metabolites such as leptin and adiponectin. leptin is a protein synthesized by adipose tissue that acts in an inhibitory way on the satiety center, thus decreasing food intake and increasing energy expenditure. It has been observed that blood leptin values increase considerably in obese dogs (although this is not the only cause of the increase). Although the values vary depending on the time elapsed since the last meal and other factors to be taken into account, table n° 1 shows the values obtained in some scientific works. Adiponectin is a cytokine synthesized only by adipose tissue, its function is to decrease plasma glucose concentration, increase fatty acid oxidation in muscle and favor the inhibition of glucose secretion by hepatocytes. Its levels decrease with obesity and in diabetic patients (7). Table n^0 1.

Table 1 Comparison of leptin and adiponectin levels in healthy and obese dogs.

Body condition	Weight (kg)	Leptin (ng/mL)	Adiponectin (µg/mL).
Healthy	10.1 (±4)	2.0 (±0-4)	37.7(±2)
Obese	13.8 (±0.6)	5.8(±0.7)	28.1(±2.3)

Other measurements

There are several more specialized clinical measurements used by veterinarians during weight loss programs in both dogs and cats for which specific measuring devices are required. These include:

• Bone densitometry or Dual X-ray Absorptiometry (DEXA): through which the precise measurement of body fat and lean tissue can be determined. - Bioelectrical impedance analysis: through which body composition is determined by measuring the conductance nature of an applied electrical current.

• Zoometry: which consists of taking a series of measurements such as hind leg and body length, including them in an equation that calculates body condition. A good example of this is the Feline Body Mass Index (BMIF). The BMIF has been designed to estimate a cat's body fat levels. This is calculated using the following equation, which uses a rib cage circumference and lower hind leg measurement in centimeters (12).

Figura 7:   Measurement of the Circumference of the thoracic cage.

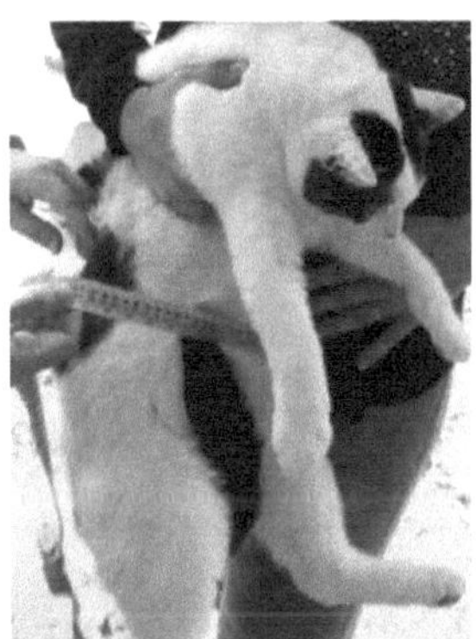

Figura 8:   Measurement of the lower leg length (LIM) is defined from the middle of the patella to the dorsal end of the calcaneal process.

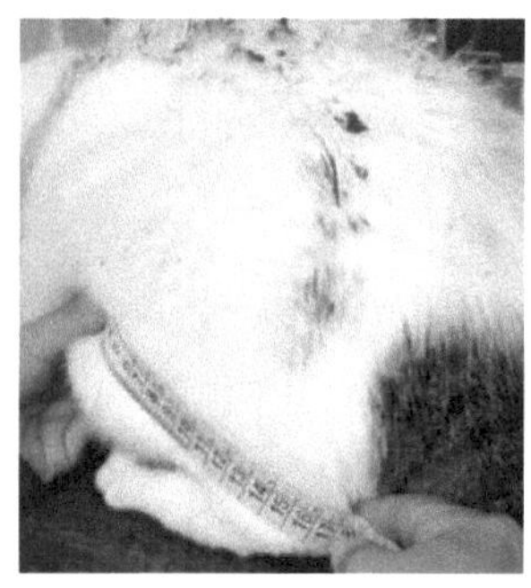

$$\text{Percentage of body fat} = \frac{[\text{caja torácica-LIM}] - \text{LIM}}{0.7062}$$

The values obtained are extrapolated to a table where the weights of the animals can be observed in order to characterize our patient.

CONCLUSIONS

Regardless of the method selected for weight control follow-up, programs should include calorie-restricted diets and exercise regimens to bring the patient to a weight that allows him/her to achieve wellness. It is also necessary to monitor compliance with weight goals on a weekly basis and to measure the patient's functional progress objectively through a daily activity monitoring system.

It is of great importance to make the owner aware of his animal's body condition since his collaboration is key, both in the rigorous diet and in the established exercises. The relationship between the individual, the environment, the clinician and the owner are key links to achieve success in the treatment.

BIBLIOGRAPHY

1. Chandler ML. Man and his canine and feline pets: ^why are we still overweight, MS, DVM, MANZCVSc, DACVN, DACVIM, DECVIM-CA,

MRCVS Vets Now Referrals Nutrition Service Espana 2015;123-145.

2. Francisco Mirò-Rodriguez C, Conde-Ruiz C., Martinez-Galisteo A. Physiotherapy: an effective means in the conservative treatment of knee osteoarthritis in dogs. RECVET, July 2007. Vol. II, N^0 07.

3. Gant P, Holden S. L., Biourge V, German A J. Can you estimate body composition in dogs from photographs? Gant et al. BMC Veterinary Research (2016) 12:18 DOI 10.1186/s12917-016-0642-7.

4. German A.J, Titcomb J.M, Holden S.L, Queau Y, Morris P.J, Biuourge V. Cohort Study of the Success of Controlled Weigth Loss Programs for Obese Dog. Journal OfVeterinary Internal Medicine. 2015; 29; 15471555.

5. Gonzalez MS, Bernal L. Diagnosis and management of obesity in dogs. A review. Rev CES Med VetZootec. 2011; Vol 6(2); 91-102.

6. Herrera Uribe J, Vitger A.D, Ritz C, Fredholm M, Bjornvad C.R., Cirera S. Physical training and weight loss in dogs lead to transcriptional changes in genes involved in the glucose-transport pathway in muscle and adipose tissues. The Veterinary Journal 2016; 208; 22-27.

7. Ishioka K, Omachi A, Sagawa M, Shibata H, Honjoh T, Kimura K, Saito M.. Canine adiponectin: Cdna structer, mRNA expression in adipose tissues and reduced plasma levels in obesity. Res. Vet. Scien. 2006; 80; 127-132.

8. Laflamme D. Development and validation of a body condition score system for dogs. Canine Pract 1997; 22: 10-15.

9. Levine, D., Millis, D., Rehabilitation of the Geriatric Patient. In:Bockstahler, B., Levine, D., Millis, D., (Eds). Essential Facts of Physiotherapy in Dogs and Cats. Rehabilitation and Pain Management. BE VetVerlag. Babenhausen. Germany.Pp. 2004; 272- 286.

10. Richardson, D.C., Schoenherr, W.D., Zicker, S.C. Nutritional magenement of osteoarthrits. Vet. Clin. North Am Small Anim Pract.1997; 27, 883-911.

11. Tvarijonaviciute A., Martinez-Subiela S., Ceron Madrigal J.J. Methods.

to measure the degree of obesity in dogs: Between physics and biochemistry. AnVet. Murcia. 2008; 24; 17-30.

12. WALTHAM® pocket book of healthy weight maintenance for cats and dogs - Factors affecting healthy weight. 2 Edition.

TREADMILL AND HYDROTHERAPY IN PROGRAMS OF CANINE AND FELINE REHABILITATION

Pellegrino FJ[1-4]

[1]Head of Practical Works, Chair of Biophysics, Department of Basic Sciences, Faculty of Veterinary Sciences (FCV), National University of La Plata (UNLP).

[2]Postdoctoral Fellow, Consejo Nacional de Investigaciones Cientificas y Técnicas (CONICET).

[3]Laboratory of Veterinary Physiotherapy (LAFIVET), FCV, UNLP.

[4]Institute of Veterinary Genetics "Ing. Fernando Noel Dulout" (IGEVET, UNLP- CONICET LA PLATA), FCV.

SUMMARY

This chapter proposes to contribute to the knowledge about the practical application of treadmill and hydrotherapy in canine rehabilitation programs, extending its potential use to the feline species. The chapter includes general concepts on the use of treadmill, the care and management of the patient during rehabilitation sessions and the different pathologies in which its use is recommended. Regarding hydrotherapy, the physical properties of water, its indications and contraindications and the use of underwater treadmill for therapeutic exercises are highlighted.

INTRODUCTION

Therapeutic exercises are a fundamental part of rehabilitation programs for companion animals (9). They are activities that help the patient to perform the correct movement of their limbs, distributing the weight evenly on them. In recent years, the use of treadmill and hydrotherapy in rehabilitation programs have become widely spread methods due to their good results. They are activities that work especially on joint movement and muscle mass strengthening. In this way, they favor locomotion, reducing claudication and pain in the patient. They also contribute to improve body condition, blood circulation and cardiorespiratory capacities. Both the use of treadmill and hydrotherapy are indicated in patients with orthopedic and neurological conditions (15, 16, 20, 22). Their use also extends to the treatment of chronic diseases and obesity in weight loss programs (6, 14, 24). Although these activities should be started when the animal is able to remain stationary, the therapist should provide assistance to the patient throughout the session.

TREADMILL REHABILITATION

The use of treadmill allows to control the exercise conditions, particularly the speed, duration and inclination of the treadmill (4, 10, 19, 21), facilitating the implementation of specific routines. For example, exercises in an inclined plane favor the recovery of muscle strength in the hind limbs and improve cardiorespiratory work, which is important in animals recovering from surgery for cranial cruciate ligament rupture and in dogs with hip dysplasia (21).

Currently, there are treadmills specifically designed for use with canines. They are longer, have side panels for protection and usually have a harness system to support the animal if necessary. Likewise, if there are no treadmills for canines, human treadmills can be adapted for this purpose (11). As for felines, although there are no treadmills designed for this species, the aforementioned can be used.

Exercises should be started gradually and with minimal intensity, when patients do not suffer from inflammation or pain. It should not be forgotten that these are patients who have undergone surgery, who have not used a limb for a long time, who are in pain, who have difficulty in walking normally, among other possible conditions. Therefore, in the first sessions the effort made by the animal should be minimal. The slow and continuous movement of the treadmill forces the patient to support the affected limb. This favors the recovery of normal locomotion. However, the patient should not be expected to distribute his weight evenly over his limbs during the first sessions, as this could cause injury to the affected limb. Also, the slow pace helps to improve coordination in the movements, proprioception, blood circulation and reduces pain.

The patient should be monitored during the session, if any alteration in gait, manifestation of pain or worsening claudication is observed, it should be stopped immediately. The therapist will be key in accompanying and monitoring the patient throughout the therapy.

In the particular case of cats it is necessary to have more patience. They are animals that are not used to daily exercise, therefore, strategies should be used to avoid that when they are stimulated to perform some type of exercise they do not feel disturbed. The environment should be calm, without stimuli of any kind that may generate discomfort or lack of concentration in the animal.

You should start with slow walks of short duration. Ideally, the walking speed should be between 1 and 2 km/h and the session time should not exceed 3 minutes. Depending on the patient's condition, the therapist can be positioned in front of the animal to stimulate walking or behind, above or on one of the patient's sides to provide support (4, 11). The patient should always be on a leash and in cases where greater assistance is needed, a harness will be placed to provide suspension and thus facilitate the patient's movements and reduce the amount of stress he/she exerts. This allows greater control during the session. The predisposition of the animal to perform the exercises should not be neglected, so that it does not lose motivation. For this it is beneficial to give rewards, encourage him, never lose sight of the logical aspect in rehabilitation programs.

The number of sessions per day should be adapted to the clinical picture, 2 or 3 sessions per day can be performed, but it will always depend on the patient we are dealing with. The animal's health should be prioritized and its evolution should be correctly evaluated when increasing the speed, duration or frequency of the sessions. In cats, improvements have been seen in spinal cord injuries in terms of locomotion and plasticity, after 3 weeks of treadmill rehabilitation (17).

Indicators in canines

- Post-surgical recovery from orthopedic injuries such as tractures (7).

- Hip dysplasia (3).

- Muscular atrophy.

- Intermittent claudication (25).

- Cranial cruciate ligament rupture (12, 15, 16).

- Chondrodystrophies (2).

- Neurological disorders causing decreased proprioception, limb tuning and loss of balance (23).

- Elbow dysplasia (20).

Indicators in felines

- Spinal cord injuries (5, 17).

HYDROTHERAPY

Hydrotherapy is the use of water as a therapeutic agent for the physical rehabilitation of patients who have suffered orthopedic and neurological injuries. It is also a useful therapy for muscle and joint strengthening. It can be performed in pools, benches or tanks in order to practice swimming exercises. Having underwater treadmills would be the ideal situation since it is possible to perform swimming exercises, standing or moving. In addition, it allows to control the workload (speed and duration), as well as the temperature and the height of the water.

In the feline species, although due to its size it could easily be carried out in any household litter box, very few cats tolerate this therapy and it is not a treatment of choice. Therefore, if it is carried out, the adaptation must be gradual and progressive. Special attention must be paid to the temperature of the water and the time of each session, as it produces muscle fatigue and physical exhaustion. The patient must always be calm, it is important to avoid excitement and stress. The height of the water will depend on the therapeutic objectives. Initially, the animal can be supported with the hands or with a support belt to help it balance and gradually gain confidence. In many situations the use of a life jacket is of great help.

Indications

Post-surgical care of orthopedic patients:

- Fracture (3).

- Osteochondritis dissecans.

- Cranial cruciate ligament rupture (1, 16).

- Hip dysplasia.

- Fore and hind limb prostheses.

Neurological pathologies or post-surgical neurological lesions:

- Disorders of intervertebral discs (8).

- Degenerative myelopathy.

- Fibrocartilaginous embolic myelopathy.

Others:

- Bilateral elbow dysplasia (20).

- Weight loss (6).

- Arthritis(3, 13).

Contraindications

- Pulmonary or cardiac conditions. Hydrostatic pressure can compress the thorax and limit its movements, mainly the inspiratory phase.

- Skin infections, mycosis, open wounds or surgical drains.

- Urinary and fecal incontinence.

- Metabolic diseases in cats such as diabetes.

Water as a therapeutic agent: physical properties

- *Relative density* : Water has a relative density of 1. Bodies with a lower density than water will float and those with a higher density will sink. Fat has a density of 0.8 (it floats) and bones have a density of 2 (they sink).

> *Flotation*: Thanks to flotation the weight in the water is reduced, so the movement of the limbs is more comfortable, reducing pain in the joints. The decrease in weight depends on the depth of immersion, the deeper the water the less weight the animal must support (Table 1). Useful in obese dogs with orthopedic disease, to avoid excessive joint load (6).

> *Hydrostatic pressure*: It is exerted homogeneously over the entire body surface. The deeper the immersion depth, the higher the hydrostatic pressure. It helps to reduce swelling and edema, improve blood and lymphatic circulation and reduce the perception of pain.

> *Surface tension*: This is determined by the forces of attraction between the water molecules, being greater on the surface than inside the liquid. This makes therapy in water easier to perform than on the surface.

> *Viscosity*: It is the resistance of a fluid to be displaced. It helps stabilize joints during movement.

> *Resistance*: Refers to the force necessary for a solid body to move through a liquid. It works on strengthening the muscles and the cardiovascular system.

Table 1. Weight supported (%) by the animal depending on the height of the water.

Water Level	Weight supported (%)
Tarsus	91

Knee	85
Hip	40

Swimming

Swimming allows coordinated muscle contraction and improves joint movements. It is an exercise that promotes blood and lymphatic circulation, improving the capabilities of the cardiovascular system (18).

It can be performed in conventional pools, swimming tanks or, depending on the size of the patient, in children's pools or home pools. Heated pools have the advantage of being able to control the water temperature.

The patient should be placed in the water without contact with the floor. The movements should be performed slowly. The use of the limbs during swimming may vary according to the dog, some use all four limbs while others use only the front limbs keeping the hind limbs flexed against the abdomen (14). As mentioned above, support should be provided with the hands, by means of a sash or with life jackets. This helps the animal feel secure and promotes balance and coordination of movements. The therapist should monitor and assist the patient throughout the therapy. If necessary, it can be placed in the water.

Underwater treadmill

Ideal for hydrotherapy. Promotes the development of muscle strength, range of motion and improves cardiorespiratory capabilities.

It allows to adapt the exercise conditions to the needs of each patient, controlling the pace and frequency. Ideally, underwater treadmills should be easily accessible for patient positioning, have ample space to facilitate movement and have suspension systems.

In principle, it is important that the animal's adaptation to this type of treatment is done gradually. Before starting any session, the animal should be warmed up by massaging, stretching and, if possible, slow walks. The first sessions are aimed at familiarizing the animal with the underwater treadmill. For this purpose, it helps a lot to play with him, to reward him, to make the use of the treadmill a pleasant exercise where the patient relaxes. The temperature of the water is another essential factor, it should be neither too cold nor too hot to avoid complications. Temperatures between 25° C and 35° C are recommended. Another important aspect is the level at which the water should be found, which is related to the patient's condition. In dogs in post-surgical rehabilitation of cranial cruciate ligament rupture, it is suggested that the water level on the treadmill should be at or above the knee (1). The patient's breathing, heart rate and mucosal status should always be monitored, and once the session is over, the animal should be dried completely.

Exercises on the underwater treadmill can be performed with the animal standing or in motion. The latter includes walking and swimming.

Standing

They are indicated in patients with neurological conditions such as degenerative myelopathy, fibrocartilaginous embolism myelopathy and post-surgical herniated discs. Thanks to the specific properties of water, patients can achieve stability, coordination of movements and gradually recover muscle strength.

It is necessary to provide support through the use of a harness or simply with the hands to facilitate stability. When the animal manages to stay in station, you can start with gentle movements of all joints and cycling movements. Another very useful exercise is body weight shifts. With the animal in station, light pressure is applied to different areas to unbalance the animal. These exercises are aimed at improving proprioception, range of motion and correct posture.

Exercised on the move

They should be done gradually, the important thing is that the animal coordinates the movements. Walking demands additional work to overcome the resistance of the water and generates fatigue. The walking speed should be slow and the session time should not exceed 2 or 3 minutes. The frequency of the sessions should be 2 or 3 times per week. Depending on the condition of the animal and its progression, in cases where you want to perform exercises with a higher degree of difficulty, you can adjust the height of the water, the speed and the inclination of the underwater treadmill.

CONCLUSIONS

Physiotherapy and rehabilitation continue to grow in veterinary medicine. Rehabilitation protocols must be adapted to each patient according to the clinical picture, its evolution and the desired goals of therapy. Hence, the treadmill or underwater treadmill and hydrotherapy are shown to be beneficial complementary methods to be used in canine and feline rehabilitation programs.

BIBLIOGRAPHY

1.   Bertocci G, Smalley C, Brown N, Bialczak K, Carroll D. Aquatic treadmill water level influence on pelvic limb kinematics in cranial cruciate ligamentdeficient dogs with surgically stabilised stifles. J Small Anim Pract. 2018; 59(2): 121-127.

2.   Blau SR, Davis LM, Gorney AM, Dohse CS, Williams KD, Lim JH, Pfitzner WG, Laber E, Sawicki GS, Olby NJ. Quantifying center of pressure variability in chondrodystrophoid dogs. Vet J. 2017; 226: 26-31.

3.   Bockstahler B, Millis D, Levine D. Indications: Classifications according to location. In: Essential facts of physiotherapy in dogs and cats: rehabilitation and pain management. Babenhausen, Germany: Be Vet Verlag; 2004. p. 126-286.

4.   Bockstahler B, Millis D, Levine D, et al. Physiotherapy-what and how. In: Essential facts of physiotherapy in dogs and cats: rehabilitation and pain management. Babenhausen, Germany: Be VetVerlag, 2004; p. 46-123.

5.   Boyce VS, Lemay MA. Modularity of endpoint force patterns evoked using intraspinal microstimulation in treadmill trained and/or neurotrophin-treated chronic spinal cats. J Neurophysiol. 2009; 101(3): 1309-1320.

6.   ChauvetA, LaclairJ, ElliottDA, GermanAJ. Incorporation of exercise, using an underwater treadmill, and active client education into a weight management program for obese dogs. Can Vet J. 2011; 52(5): 491 -496.

7.   Doyle ND. Rehabilitation of fractures in small animals: Maximize outcomes, minimize complications. Clin Tech Small Anim Pract. 2004; 19(3): 180-191.

8.   Drum MG. Physical rehabilitation of the canine neurologic patient. Vet Clin North Am Small Anim Pract. 2010; 40: 181-193.

9.   Drum MG, Marcellin-Little DJ, Davis MS. Principles and applications of therapeutic exercises for small animals. Vet Clin North Am Small Anim Pract. 2015; 45(1): 73-90.

10. Ferasin L, Marcora S. Reliability of an incremental exercise test to evaluate acute blood lactate, heart rate and body temperature responses in Labrador retrievers. J Comp Physiol B. 2009; 179(7): 839-845.

11. Hamilton S, Millis DL, Taylor RA, Levine D. Therapeutic exercises. In: Millis D, Levine D, Taylor RA, Editors, Canine Rehabilitation and Physical Therapy. Philadelphia: Saunders Elsevier. 2004; p. 244-263.

12. Jerre S. Rehabilitation after extra-articular stabilization of cranial cruciate ligament rupture in dogs. Vet Comp Orthop Traumatol. 2009; 22(2): 148152.

13. Levine D, Rittenberry L, Millis DL. Aquatic therapy. In: Millis D, Levine D, Taylor RA, Editors. Canine Rehabilitation and Physical Therapy. Philadelphia: Saunders Elsevier, 2004; p. 264-276.

14. Marcellin-Little DJ. Therapeutic exercise. In: Towell TL, Editor. Guia practica para el control de peso de perros y gatos. Bogota, Colombia: Editorial Inter-Médica. 2013; p. 143-163.

15. Marsolais GS, Dvorak G, Conzemius MG. Effects of postoperative rehabilitation on limb function after cranial cruciate ligament repair in dogs. JAVMA. 2002; 220(9): 1325-1330.

16. Marsolais GS, McLean S, Derrick T, et al. Kinematic analysis of the hind limb during swimming and walking in healthy dogs and dogs with surgically corrected cranial ligament rupture. JAVMA. 2003; 222(6): 739-743.

17. Martinez M, Delivet-Mongrain H, Rossignol S. Treadmill training promotes spinal changes leading to locomotor recovery after partial spinal cord injury in cats. J Neurophysiol. 2013; 109: 2909-2922.

18. Mikail S, Pedro CR. Veterinary Physiotherapy. Brazil: Manole Ltda. 2006.

19. Pellegrino FJ, Risso A, Relling AE, Blanco PG, Arias DO, Corrada Y. Effect oftreadmill training on cardiac size, heart rate and muscle mass in healthy dogs. J Vet Adv. 2014; 4(9): 686-690.

20. Preston T, Wills AP. A single hydrotherapy session increases range of motion and stride length in Labrador retrievers diagnosed with elbow dysplasia. Vet J. 2018; 234: 105-110.

21. Saunders DG. Therapeutic exercise. Clin Tech Small Anim Pract. 2007; 22(4): 155-159.

22. Shmalberg J, Memon MA. A retrospective analysis of 5,195 patient treatment sessions in an integrative veterinary medicine service: patient characteristics, presenting complaints, and therapeutic interventions. Vet Med Int. 2015.

23. Sulka KA, Christy MR, Peterson WL, Rudd SL, Troy SM. Reduction of pain- related behaviors with either cold or heat treatment in an animal model of acute arthritis. Arch Phys Med Rehabil 1999; 80(3):313-317.

24. Vitger AD, Stallknecht BM, Nielsen DH, Bjornvad CR. Integration of a physical training program in a weight loss plan for overweight pet dogs. J Am Vet Med Assoc. 2016; 248(2): 174-182.

25. Weiss T, Fujita Y, Kreimeier U, Messmer K. Effect of intensive walking exercise on skeletal muscle blood flow in intermittent claudication. Angiology. 1992; 43(1): 63-7